Praise for *Pictures of the Mind*

"*Pictures of the Mind* is an extraordinary book. It makes the unfolding scientific story of consciousness vivid, even joyous, while offering a sophisticated tour of what is known about ourselves, our emotions, and our brains. A beautiful read."

—Ruth R. Faden,
Philip Franklin Wagley Professor of Biomedical Ethics and
Director of the Berman Institute of Bioethics at Johns Hopkins University

"This book explores the many ways in which neuroscience is revealing remarkable things about the inner workings of our minds—not the least of which is the transformative impact that meditation can have on destructive thoughts and behavior. I have no doubt that Miriam Boleyn-Fitzgerald's work will be of great benefit to those with an interest in this fascinating new area of inquiry."

—Yongey Mingyur Rinpoche,
Author of *The Joy of Living: Unlocking the Secret and Science of Happiness*

"The mind is embodied, and it is relational. In this straightforward and illuminating book, Miriam Boleyn-Fitzgerald reveals pictures in both visual and narrative form that capture the power of the mind to transform the brain. How our mental lives shape and are shaped by neural circuitry—itself forever being molded by experience—is the central theme of these powerful portraits of what it means to be human. By learning to focus our minds in more compassionate ways—toward ourselves and others—we can literally promote a healthier and more integrated brain. Read these pages, and you'll be able to see for yourself!"

—Daniel J. Siegel, M.D.,
Author of *Mindsight: The New Science of Personal Transformation*,
Clinical Professor of Psychiatry at the UCLA School of Medicine,
Co-Investigator at the Center for Culture, Brain, and Development, and
Co-Director of the Mindful Awareness Research Center

"Miriam Boleyn-Fitzgerald has given us a remarkably clear and engaging account of the ways that the new brain imaging technologies can give us deep insights into our gravest maladies. Her conclusion, that healing may often lie with us, joins science with the wisdom of the ages."

—Jonathan D. Moreno,
Author of *Mind Wars*, David and Lyn Silfen University Professor, and
Professor of Medical Ethics and of History and
Sociology of Science at the University of Pennsylvania

"An engaging and compelling read that illustrates how the new brain science can help us understand elements of our basic humanity."

—Zindel Segal,
Author of *The Mindful Way through Depression* and
Cameron Wilson Chair in Depression Studies at the University of Toronto
and the Centre for Addiction and Mental Health

Pictures of the Mind

Pictures of the Mind

*What the New Neuroscience Tells Us About
Who We Are*

Miriam Boleyn-Fitzgerald

Vice President, Publisher: Tim Moore
Associate Publisher and Director of Marketing: Amy Neidlinger
Editorial Assistant: Pamela Boland
Development Editor: Kirk Jensen
Operations Manager: Gina Kanouse
Senior Marketing Manager: Julie Phifer
Publicity Manager: Laura Czaja
Assistant Marketing Manager: Megan Colvin
Cover Designer: Stauber Design Studio
Managing Editor: Kristy Hart
Project Editor: Jovana San Nicolas-Shirley
Copy Editor: Deadline Driven Publishing
Proofreader: Seth Kerney
Senior Indexer: Cheryl Lenser
Compositor: Jake McFarland
Manufacturing Buyer: Dan Uhrig

Cover image courtesy of Steven Laureys, from S. Laureys, "The Neural Correlates of (Un)awareness: Lessons from the Vegetative State," *Trends in Cognitive Sciences* 9, no. 12 (December 2005): 556-559, http://www.sciencedirect.com/science/journal/13646613. Copyright 2005, reprinted with permission from Elsevier.

© 2010 by Pearson Education, Inc.
Publishing as FT Press
Upper Saddle River, New Jersey 07458

FT Press offers excellent discounts on this book when ordered in quantity for bulk purchases or special sales. For more information, please contact U.S. Corporate and Government Sales, 1-800-382-3419, corpsales@pearsontechgroup.com. For sales outside the U.S., please contact International Sales at international@pearson.com.

Company and product names mentioned herein are the trademarks or registered trademarks of their respective owners.

Printed in the United States of America

First Printing January 2010

ISBN-10: 0-13-715516-6
ISBN-13: 978-0-13-715516-3

Pearson Education Ltd.
Pearson Education Australia PTY, Ltd.
Pearson Education Singapore, Pte. Ltd.
Pearson Education North Asia, Ltd.
Pearson Education Canada, Ltd.
Pearson Educación de Mexico, S.A. de C.V.
Pearson Education—Japan
Pearson Education Malaysia, Pte. Ltd.

Library of Congress Cataloging-in-Publication Data

Boleyn-Fitzgerald, Miriam.
    Pictures of the mind : what the new neuroscience tells us about who we are / Miriam Boleyn-Fitzgerald.
        p. cm.
    ISBN-13: 978-0-13-715516-3 (hardback : alk. paper)
    ISBN-10: 0-13-715516-6 (hardback : alk. paper) 1. Neurosciences. 2. Brain--Psychophysiology.
I. Title.
    RC341.B65 2010
    616.8--dc22
                                                                                                    2009040031

In memory of my brother Steven.

Contents

Introduction

Imagine your mind is a yard on a clear March day. You've been offered a chance to walk around. You may choose to clear fresh paths through the brown winter muck, pick up bits of trash you forgot were buried under all that ice and snow, decide what new seeds to plant and where. You've been told your soil is rich—much richer than you thought—and now you are sure that with time and attention and a lot of gritty work, you can grow almost anything. You roll up your sleeves and take stock. What habits of mind will you dig up and toss on the compost heap? What mental skills and emotional states, what beliefs about yourself and the world will you choose to cultivate?

Little more than a decade ago, the physical landscape of our minds was perfectly invisible to us and, for all we knew, as fertile and productive as it was ever going to get. Even if we were aware of our thoughts, ideas, and emotions, we had no way of watching the neural activity associated with mental phenomena arise, do its thing inside our heads, and pass away. We had no way of watching that activity actually alter and strengthen our neural networks. Now, thanks to powerful new imaging tools like functional MRI (fMRI) and positron emission tomography (PET), we can watch the organ of the mind in action, and what we see is exhilarating: The brain has the capacity to heal, grow, and change itself in ways that before were thought impossible.

Conventional scientific wisdom used to paint a starkly different picture of the adult brain, one in which its physical structure was essentially immutable. By the age of three, the story went, most neural networks were in place, and by late adolescence, our temperament—our baseline chemical state of happiness or irritability—was thought to be fixed. If we had always been a sunny kid, our outlook would probably tend toward the golden for the rest of our lives, but for those of us born and raised on the dark side of the moon, we would probably always struggle with negative emotions like anxiety, sadness, and aggression. Compounding the depressing picture was the conviction that, were we to lose nerve cells through disease, aging, or injury, there was very little point in ever wishing them back.

Now we can watch our brains on-screen, healing and adapting to challenges, and we see that our genes and early experiences absolutely *do* influence our cognitive and emotional makeup in important ways, and that they absolutely *don't* get to dictate who we become. This previously unappreciated flexibility and trainability of neural pathways is termed "neuroplasticity," and it has transformed modern neuroscience into an intensely optimistic field where researchers seek new diagnostic techniques and therapies for patients recovering from structural damage and chemical imbalances due to traumatic brain injury, stroke, Alzheimer's disease, emotional disorders, drug addiction, and chronic pain. Research into neuroplasticity isn't just revealing how we can heal brain injuries and sharpen our wits, but also how we can strengthen key neural pathways to become happier, kinder, less fearful, and more effectual merely by changing the way we perceive the world and our reactions to it.

The idea that humans are especially resilient creatures isn't new. Aristotle, for one, thought that the brain's chief biological role was to cool the heart, which he took to be the true physical seat of thought, reason, and emotion. The brain "tempers the heat and seething of the heart,"[1] the great philosopher wrote, and because "the heat in man's heart is purest,"[2] humans require much bigger brains than beasts. We are the wisest, the best tempered of animals, he conjectured, because our brains are big enough to cool our hot, hot blood.

Our knowledge about the brain's regulatory role has evolved since Aristotle's time, but there is a sense in which he got it exactly right. Images of the brain recovering from emotional disorders, traumatic memories, and addiction show that this remarkable organ can, in fact, "cool the heart." The self-regulating brain can alleviate the intense physiological effects of primal emotions like anger and fear, and in doing so, it can heal emotional damage and protect the body. We may even one day learn to cure the excruciating cravings that keep many of us locked in cycles of addiction by consciously modulating activity in particular areas of the brain. It seems that we've always instinctually known that we are born to be highly adaptable—that we are gifted with extraordinary abilities to heal and change in the face of adversity—and now we are taking the pictures to prove it.

Cutting-edge imaging research also holds the promise of providing new diagnostic and therapeutic options for patients suffering from states of impaired awareness or—most harrowingly—from awareness without any ability to communicate with the outside world. A patient's thoughts might be observed on-screen in ways that could be understood by others—an exciting possibility for families and friends who spend months or even years at the bedside of a loved one recovering from a brain injury. At one heartbreaking extreme, of course, are the patients who never recover any meaningful level of awareness. Stem-cell researchers are working hard to grow functioning neurons in the laboratory, but until such research bears therapeutic fruit, there are certain severe brain injuries—global neuronal loss due to prolonged oxygen deprivation, for example—that simply cannot be repaired.

Standard bedside tests for consciousness have remained largely unchanged for 30 years, as have the painful ambiguities that can accompany end-of-life decisions. Brain-imaging technologies may ease the agony of some of these complex choices by enhancing our understanding of the neural correlates of awareness. Pictures of the brain responding to the world may one day become a standard tool to clarify when medical interventions are working to extend meaningful life and when they are inappropriately and painfully prolonging death.

New imaging tools also mean new hope for age-related cognitive decline. Functional MRI, PET, and a revolutionary imaging substance called Pittsburgh Compound-B (PiB) have revealed that Alzheimer's disease attacks different parts of the brain than those affected by normal aging, and that some brain systems, including some forms of memory, are left intact by Alzheimer's. It is possible that these undamaged systems may be trained to help people afflicted with the disease to function better. And for those of us with a few million miles on the brain, there is more to be excited about: Research into new diagnostic methods may help catch the debilitating disease in its earliest stages (when drugs and other forms of therapy might be most effective), while other studies show that certain kinds of mental training can relieve the effects of "normal" age-related memory decline.

Tantalizing pictures of the brain are also emerging from the field of neurotheology, a branch of learning that seeks connections between spiritual experiences and mental activity. Neuroscientists have scanned the brains of spiritual practitioners as diverse as Tibetan Buddhists and Franciscan nuns, and have found striking similarities in their brain activity when they are in states of "higher consciousness"—states in which people stop sensing a separation between themselves and the world, in which their minds feel limitless, expansive, and in touch with God or spiritual insight. Researchers are also identifying neural pathways for spiritually significant mind states like empathy, compassion, and forgiveness, showing not only that prosocial emotions are skills that can be perfected through training, but that practicing them makes our brain circuitry more positive and responsive in ways that could be used to prevent depression and other mood disorders, bullying and violence, and even physical damage inflicted by our own negative internal states.

Pictures of the Mind looks at images across the full spectrum of consciousness—from impaired, to healthy, to "higher"—and at what they tell us about the brain's extraordinary capacity to heal after illness and injury, to adapt to new challenges, and to retrain itself in ways that can make us happier, healthier people, more attentive to our own needs and to the needs of others. This deepening knowledge about the organ of knowledge is transforming our basic understanding of the "self" from a static and constricted entity to a vast and productive landscape—the ideal ground in which to cultivate conditions for well-being.

Part I

Snapshots

1

Life, death, and the middle ground

The last thing you feel is the brute force of a head-on collision. The paramedics arrive at the scene to find you unconscious, a victim of massive head trauma in a devastating car accident. They rush you to the nearest emergency room, where doctors do everything they can to stem the bleeding and limit the damage to your brain.

Your loved ones arrive at the hospital, and they are told to prepare for the worst. Only time and a battery of neurological tests will reveal the true nature and severity of your injuries. You spend several days in a coma, and then, to the great relief of your family and friends, you open your eyes. Everyone close to you rejoices. They take this development as a sign of recovery—that you can see them, hear them, know that they love you. But the doctors caution that it is too early to know if you will ever recover awareness of your surroundings. Their bedside tests indicate that so far, you have not.

You wake up and fall asleep just like a healthy person. This encourages the people who love you, but it does not prove awareness. Sleep cycles are controlled by your brain stem—the most primitive part of your brain—not by regions involved in conscious perception or thought. Over the next several months, you consistently fail standard neurological tests designed to establish whether you are a thinking, feeling, and aware person. You show no overt behaviors that could be read as willed, voluntary, or responsive. Your doctors see no reason for hope. After five months, they diagnose you as persistently vegetative.

Your case, it turns out, is extraordinary, and—quite harrowingly—you are the only person who knows it. You are, in fact, fully aware of yourself and your surroundings, but you are incapable of making any purposeful movements at all. You cannot even blink your eyes on command. It is the stuff of nightmares—trapped inside a body that you cannot control. Without tools to communicate your thoughts or feelings to loved ones, doctors, or nurses, you feel utterly terrified and alone, cut off from the world. You have no way of knowing whether anyone will ever detect that your mind is still alive, feeling the pain of total isolation.

Can such a catastrophic fate exist? It can, and it has in rare cases, say researchers who are working to scan the brains of patients diagnosed as vegetative.

In July 2005, a 23-year-old English woman was critically injured in a car accident. For a full five months she failed clinical tests of consciousness and her doctors declared her vegetative. Not until half a year after her accident—when researchers in Cambridge, England selected her to take part in a study of brain activity in vegetative patients using state-of-the-art functional magnetic resonance imaging (fMRI)—was her surprising brain activity discovered.

Picturing awareness

MRI technology uses an extraordinarily powerful magnetic field (often up to 30,000 times stronger than the earth's magnetic field) to align excited water molecules with—or against—the direction of the force. The water molecules absorb or transmit radio waves, producing a pattern detected and analyzed by a computer. Structural MRIs have been producing high-resolution, two- and three-dimensional images of the brain since the 1980s.

Functional MRI scanning, first developed in the early 1990s, has been refined and applied at an astonishing pace. The technology uses an MR signal to measure blood-flow changes by recording shifts in blood-oxygen levels. When activity in a certain part of the brain increases, so does the need for fresh oxygen. Blood rushes in, causing the MR signal to increase. In this way, fMRI technology can reveal which parts of the brain are working and under what circumstances.

The young English patient was placed inside an fMRI scanner, and something remarkable happened. When she heard spoken sentences, and then acoustically matched but meaningless noise sequences, her brain was able to tell them apart, lighting up in telltale language-processing patterns in response to the meaningful sentences.

Lead investigator Adrian M. Owen and the rest of the British and Belgian researchers studying her case were careful to point out that these patterns by themselves were not indisputable evidence of awareness. Studies of "implicit learning and priming," they emphasized in their report, "as well as studies of learning during anesthesia and sleep, have demonstrated that aspects of human cognition, including speech perception and semantic processing, can go on in the absence of conscious awareness."[1]

But then the truly startling result: Asked to respond to mental imagery commands—first, imagining herself playing tennis, and then picturing herself walking through the rooms of her home—her brain responded instantly and sustained the mental "work" for a full 30 seconds. On screen, her patterns of mental activity were measured by blood traffic to movement and imagery centers in her brain and looked just like a healthy person's.

This extraordinary result led the neuroscientists studying her case to conclude that she was "beyond any doubt...consciously aware of herself and her surroundings." Next to brain scans of 12 healthy volunteers, reported Dr. Owen, "you cannot tell which is the patient's."[2] And six months later—one year post-accident—she was able to follow a mirror intermittently to one side, thus joining a relatively new category of patients described as "minimally conscious."

Dr. Owen and his co-investigators have faced intense criticism from fellow neuroscientists who believe that the findings were overstated—that the fMRI results cannot be deemed clear evidence of purposeful mental activity. The crux of detractors' arguments is the idea that words presented to the patient in the imagery tasks—words like "tennis" and "house"—could have triggered her mental response even in the absence of awareness.

Not so, said Owen and colleagues in their spring 2007 response in the journal *Science*. Such automatic changes in the absence of conscious awareness "are typically transient (i.e., lasting for a few seconds) and, unsurprisingly, occur in regions of the brain that are associated with word processing. In our patient, the observed activity was not transient but persisted for the full 30 [seconds] of each imagery task.... In fact, these task-specific changes persisted until the patient was cued with another stimulus indicating that she should rest. Such responses are impossible to explain in terms of automatic brain processes." What is more, her responses were observed not in the word-processing centers—as would be expected in an unconscious patient—but in "regions that are known to be involved in the two imagery tasks that she was asked to carry out."[3]

Psychologist Daniel Greenberg proposed the following test of whether the patient had made a conscious decision to cooperate: What would happen, he asked, if investigators "presented a similar noninstructive sentence such as 'Sharleen was playing tennis?'"[4] Owen and colleagues addressed Greenberg's question by prompting a healthy volunteer with these sorts of noninstructive statements, and the results were persuasive: No activity was observed in any of the brain regions that had been triggered in the patient or healthy volunteers when they were performing the mental imagery tasks. This result, the researchers argued, reaffirmed their original conclusion that the patient was knowingly following instructions, despite her diagnosis to the contrary.

A rare mistake?

The implications of the discovery are haunting. How many patients currently labeled vegetative might show similar patterns of mental activity on an fMRI scan? There are an estimated 25,000 to 35,000 vegetative patients in the United States alone.[5]

Nicholas Schiff, a leading American researcher on impaired consciousness, says that while the study showed "knock-down, drag-out" proof of awareness in this patient, it is unclear "whether we'll see this in one out of 100 vegetative patients, or one out of 1,000, or ever again."[6] Most neuroscientists are of the opinion that the vast majority of vegetative cases would not show the complex mental activity that the young English woman did. Many of these patients—Theresa Schiavo among them—have suffered extreme oxygen deprivation and have lost a massive number of neurons. Steven Laureys, director of the Coma Science Group at the Cyclotron Research Center at the University of Liége in Belgium, reports that his team has "not observed any similar signs of awareness in functional scans of more than 60 other vegetative patients studied at the University of Liége."[7]

Dr. Laureys coauthored the imaging studies with the young English patient, and he notes that even for experts in impaired consciousness, "the vegetative state is a very disturbing condition. It illustrates how the two main components of consciousness can become completely dissociated: wakefulness remains intact, but awareness—encompassing all thoughts and feelings—is abolished."[8]

The chance of recovery is greater for victims of a traumatic brain injury like the young English patient, whose injuries are severe but localized. An estimated one-half of all patients who are unconscious due to traumatic injuries—often the victims of car accidents—regain some awareness within a year, whereas only about 15 percent of those with brain damage due to oxygen starvation recover any awareness within the first three months (and very few do after that). A study of 700 vegetative patients in 1994 showed that no patients with injuries from oxygen deprivation recovered after two years.[9]

"One always hesitates to make a lot out of a single case, but what this study shows me is that there may be more going on in terms of patients' self-awareness than we can learn at the bedside," said Dr. James Bernat, professor of neurology at Dartmouth Medical School. "Even though we might assume some patients are not aware, I think we should always talk to them, always explain what's going on, always make them comfortable, because maybe they are there, inside, aware of everything."[10]

Kate Bainbridge from Cambridgeshire, England, agrees. She believes that she, too, has much to owe Dr. Owen's brain imaging research. She was 26 years old when she succumbed to a viral infection that caused severe brain inflammation, leaving her vegetative for six months according to standard bedside tests. Dr. Owen had begun using scanning technology—in Kate's case, positron emission tomography (PET)—to investigate patterns of brain activity in patients diagnosed as vegetative. A PET scan takes a biologically active chemical like glucose, tags it with a detectable radioisotope, and records its uptake by active brain cells.

Dr. Owen showed Ms. Bainbridge photographs of her family, and he saw that she recognized them when facial-processing centers in her brain ignited on-screen. When she was shown nonsense images with similar colors, these regions of her brain lay dormant. Dr. Owen performed additional PET scans over the coming months, and they showed that she was becoming more aware all the time.

"Not being able to communicate was awful—I felt trapped inside my body," Kate said. "I had loads of questions, like 'Where am I?', 'Why am I here?', 'What has happened?' But I could not ask anyone—I had to work it all out."[11] She formed nightmarish theories to try to make sense of it all. "I thought I was in prison and I had forgotten how to move."[12]

But the PET scans changed all that, she said. "They found I was there inside my body that did not respond."[13] She credits the scans for hastening her recovery, for giving her hope, and for encouraging others to interact with her. It took two years before she regained full consciousness. Now she uses a wheelchair, but she can communicate using a keyboard and she can read, use the computer, and play board games without help.

Her parents agree that Dr. Owen's discovery was a critical piece of news. "The scan meant an enormous amount to us," her father said. After the test, "the doctors were able to tell us for the first time that Kate's brain was processing things. That was a big breakthrough and it meant when we were doing things with her—talking to her, showing her pictures, writing her notes—we felt, even if she didn't understand, her brain was processing things. We realized something might be there to help her to cope with this horrendous experience she was going through."[14]

When she was told the news that the Cambridge team had discovered what was apparently normal brain activity in another "vegetative" patient, Ms. Bainbridge observed, "I think the work Dr. Owen is doing is so important. I can remember how awful it was to be like I was. It really scares me to think what could have happened if I hadn't had the scan."[15]

When hope is in vain

The journal *Science*, when it published the remarkable fMRI findings, cautioned that they must not be taken as typical. The editors included a special note in their press release about the study, mentioning the controversial Theresa Schiavo case and stressing that the new research "should not be used to generalize about all other patients in a vegetative state, particularly since each case may involve a different type of injury."[16]

What about Theresa Schiavo and her heart-wrenching case? Few end-of-life stories have polarized a country the way hers did, and some who followed it might wonder if the results of this new study call her diagnosis as permanently vegetative into question. Was there ever any reasonable hope for Theresa's recovery?

Decidedly not, say neuroscientists asked to comment, noting that her brain was deprived of oxygen for much too long. By the time paramedics arrived to resuscitate her heart and breathing, her brain was already severely, irreparably damaged. All seven board-certified neurologists who examined her agreed that she was permanently vegetative—that the thinking, feeling part of her brain had died. "As such," says neurologist Eelco Wijdicks of the Mayo Clinic College of Medicine, "she could never have recovered to an independently functioning human being, able to care for herself."[17]

Despite consensus among neurologists who examined her, Ms. Schiavo's health was described a number of ways—many of them erroneous—as the public debate raged around removal of her life support. Some neurologists near the end of her life suggested that she might be minimally conscious, whereas other commentators described her as brain dead or, at the other end of the cognitive spectrum, as a victim of locked-in syndrome (LIS), a condition in which there is severe damage to the brainstem but the cerebral cortex is unaffected. A person with LIS can think and feel emotion just like a healthy person, but cannot move or communicate except by blinking.

What role did brain scans play in her diagnosis? Dr. Ronald Cranford, who examined Theresa in 2002, performed a computed tomography (CT) scan of her brain. CT technology takes a series of two-dimensional x-ray images and compiles them, through computer algorithms, into three-dimensional images.* Dr. Cranford reported

---

* The CT scanner's existence is thanks in part to the British musical group The Beatles, as their impressive record sales enabled their label, EMI Music Publishing, to fund the research of one of its inventors.

that the scan showed little else than scar tissue and spinal fluid. He also performed an electroencephalogram (EEG), a diagnostic test to record electrical impulses (brain waves) associated with mental activity. The EEG showed no sign of life in the thinking parts of her brain.

"It's totally flat—nothing," Dr. Cranford said, "and this is very unusual. The vast majority of people in a persistent vegetative state show about 5 percent of normal brain activity."[18]

The autopsy report, released in June 2005, backed the neurologists' diagnosis. "This damage was irreversible," the medical examiner said of the injuries to her brain, which had shrunk to half its normal size. "No amount of therapy or treatment would have regenerated the massive loss of neurons."[19]

A spectrum of awareness

Even if the young English woman's story proves a rare case, it has prompted concern over whether standard bedside tests for the vegetative state are reliable and whether the boundaries between states of consciousness are clear at all.

The persistent vegetative state (PVS), as we currently understand it, lies on a continuum of impaired brain function. Though neither comatose nor brain dead, vegetative patients are unaware of themselves and their surroundings; they are alive without consciousness. They can wake and fall asleep, but they cannot communicate or respond to commands in a meaningful way. The thinking, feeling part of the brain (the cerebral cortex) no longer functions, but the more primitive part of the brain governing reflexes (the brainstem) still operates. Isolated areas in the cortex may still show activity, but they are disconnected from parts of the brain necessary for conscious perception.

"There are islands of activity in a sea of silence,"[20] says Steven Laureys, who has taken PET scans of diverse states of awareness in the hopes of identifying regions in the cerebral cortex—and connections among these regions and other parts of the brain—that may prove crucial to maintaining the experience of consciousness.

Vegetative patients like Theresa Schiavo are often mislabeled as comatose or brain dead, though their brainstems are still fully operative, allowing their hearts and lungs to work and producing sleep-wake cycles. A comatose patient lingers in a state of profound unconsciousness from which she cannot be roused—even by powerful stimulation—whereas brain death entails total and permanent loss of all brain function and is one of the medical and legal definitions of death. Patients who are brain dead can no longer breathe for themselves, and their hearts will stop beating if they do not receive oxygen from mechanical ventilation. Despite these significant distinctions, a surprising 1996 survey published in the *Annals of Internal Medicine* revealed that almost half of U.S. neurologists and nursing home medical directors believed that patients in a vegetative state could be declared dead.[21]

Current diagnostic guidelines allow patients to be declared permanently vegetative after one year if they have suffered traumatic brain injury (like the young English patient) or after six months if they have suffered brain injury due to oxygen deprivation (like Theresa Schiavo). Rarely, people with oxygen-related brain damage have regained consciousness after being diagnosed as permanently vegetative, but all of those patients recovered within two years. In a few astonishing cases, vegetative patients with traumatic brain injuries have regained consciousness much later. Terry Wallis, an Arkansas mechanic, recovered awareness in 2003, more than 18 years after a serious car accident. Mr. Wallis, many neurologists now believe, would have been more accurately described as "minimally conscious."

A new class of consciousness

The past few years have seen the emergence of a controversial new category of impaired consciousness, the minimally conscious state (MCS). The term was first used in 2002 to describe people who previously would have been diagnosed as vegetative, but who can track movement with their eyes and may respond intermittently. Eelco Wijdicks of the Mayo Clinic describes MCS as "the most severe form of neurological disability in a conscious patient."[22]

Neurologists agree that it is vital to be able to distinguish between MCS and PVS in a patient, though many are still uncertain if MCS covers a single condition or a wide range of disorders. The diagnostic criteria are "difficult to define," says Wijdicks, "and the boundaries are uncertain (how minimal and how maximal?)."[23]

An estimated 100,000 Americans exist in this state of intermittent awareness, and some do recover fully. "It took years to get some agreement on the definition," says neurologist Nancy Childs of the Healthcare Rehabilitation Center in Austin, Texas, "and it's only now getting some acceptance, but we've known for years that there was this other group."[24] In the early 1990s, studies designed by Dr. Childs and Dr. Keith Andrews of London's Royal Hospital for Neurodisability discovered signs of awareness in more than one-third of patients who had originally been diagnosed as vegetative.

Brain scans have figured highly in defining the new diagnosis. A landmark study by Joy Hirsch of Columbia University and Nicholas Schiff of Cornell University's Weill Medical College used fMRI scans of two minimally conscious patients to examine neural characteristics of the state. During the scan, the investigators played recordings of close family members talking about familiar events in the patients' lives.[25] What Hirsch and Schiff found shocked them: The minimally conscious patients showed mental activity similar to healthy volunteers in response to the meaningful stories—though they showed less activity compared to healthy brains when narratives were played in reverse (and thus were linguistically meaningless).

"It was haunting,"[26] Hirsch recalled when she discussed the findings. Results like these, though not ironclad proof of awareness, suggest that there may be cognitive life in patients who cannot respond to simple commands or communicate reliably.

Silent witness

Nick Chisholm was a gifted 23-year-old athlete—in love with sports and life—when he took a devastating hit on a New Zealand rugby field. His vision went blurry and he felt sick immediately, but he chalked his symptoms up to a minor concussion. When he came off the field, he asked his coach to put him back in after a mere ten minutes' rest—but then he collapsed and paramedics rushed him to the hospital.

Three days later, it was still unclear what had happened to Nick, but he seemed to be recovering well. The doctors were going to let him go home. Then he nearly collapsed again in the hospital shower, and for the next six days suffered a series of seizures that ultimately left him paralyzed and unable to speak. A battery of tests—which felt like "all the tests known to man,"[27] as Nick reports it—revealed that he had suffered several strokes due to a dissection of his vertebral arteries. One of the strokes was so massive that it effectively annihilated his brain stem, severing the connections between his higher brain and the rest of his body.

After that horrific accident in 2000, Nick was diagnosed with locked-in syndrome (LIS), an extremely rare condition in which a person's conscious mental life and senses remain intact, but in which he is unable to move or to use his body to communicate (except by shifting his eyes or by blinking). LIS can result from a sudden injury like Nick's, or it can be due to progressive degeneration of motor neurons in the devastating neurological condition amyotrophic lateral sclerosis (ALS).

It took some time—an excruciating stretch for Nick—before he was properly diagnosed. The doctors didn't know that he was aware and listening when they told his mother that he would probably die, and then offered her the option of withdrawing life support. It was Nick's family that guessed that he wasn't comatose at all, but instead was fully conscious and able to follow every word that was said. When asked by TV New Zealand several years later to share the most frightening thing he heard during that time, Nick did not hesitate to answer that it was the doctors speaking of turning off his life support machine.[28]

"Nick's mother and his girlfriend pleaded with the medical staff to realize that he was aware of what was happening,"[29] reports medical ethicist Grant Gillett, who co-wrote a piece with Chisholm for the British medical journal *BMJ*. "When the clinicians appreciated that the diagnosis was locked-in syndrome, the climate of care changed."

Misdiagnosis of LIS is tragically common, as it is often mistaken for the vegetative state or even coma. As happened with Nick, more than half of LIS cases are identified by family members rather than by physicians or nurses, because family members tend naturally to be more attuned to the patient's needs and to read signs of a conscious presence. A 2002 survey of 44 LIS patients found that diagnosis took an average of two-and-a-half months,[30] and in some harrowing cases, it took as long as six years. Even after family members catch on, they may still have trouble convincing the attending physicians that their loved one is fully conscious.

In recent years, brain scans have shed terrifying light on the acute experience of LIS—what it is like to realize that you are conscious inside an unresponsive body without the means to communicate your experience to others. PET scans have shown that glucose metabolism in the brain's higher regions does not differ significantly in patients with LIS as compared to healthy age-matched controls, supporting the conclusion that their injury is restricted to physical paralysis, and that these patients can, as Laureys and colleagues report, "recover an entirely intact intellectual capacity."[31] PET images have also revealed a telling neural signature in acute LIS patients, one that differs strikingly from healthy controls: The amygdala—a primitive part of the brain linked to primal emotions like fear and anxiety—was hyperactive in acute, but not in chronic, LIS patients.

"It is difficult to make judgments about patients' thoughts and feelings when they awake from their coma in a motionless shell," say Laureys and colleagues, but in light of the evidently normal metabolism in higher regions of the brain, these researchers hypothesize that the pronounced amygdala activity in conscious patients who had not yet learned to communicate with the outside world "relates to the terrifying situation of an intact awareness in a sensitive being, experiencing frustration, stress, and anguish, locked in an immobile body." They conclude from this preliminary imaging evidence that medical professionals caring for patients with LIS must adjust their bedside behavior to address states of extreme emotional distress.

We can only imagine, Gillett notes, "the sheer awfulness of hearing others discuss turning off one's life support."[32] Even after medical professionals knew that Nick was fully conscious and aware of everything unfolding in the room, Gillett says that Nick still "heard things said about his prognosis and his life that paid little regard to what he might have been thinking." Nick reports that his case manager for New Zealand's Accident Compensation Corporation said in his presence that even if he did live, he wouldn't want to anyway, and a specialist told him to get used to the wheelchair, because he would be in one for the rest of his life. In his second year in the hospital, another specialist told him that whatever gains he had made to date were the only improvements he'd ever see.

Nick begged to differ. "What do they really know? They only know what they read in textbooks," he says, and he characterizes most of his specialists and doctors as "so extremely negative."

In cases of locked-in syndrome, clinicians naturally make mistakes all the time about what a patient is thinking or feeling, Gillett says, because it is a state of consciousness that simply defies our ability to know how it feels to be the other person. "We must make special efforts to reach past the locked-in syndrome barrier and allow the patient to connect with us," he argues. No matter how closely Nick's condition resembles other states devoid of awareness, he must be treated like a whole person, capable of experiencing a complete range of thoughts, feelings, and emotions.

Rather than experiencing cognitive deficits, Nick feels that some of his senses—sight and hearing in particular—have actually been enhanced by his situation. Because of the natural tendencies of people surrounding LIS patients to dwell on everything that has been lost, it is all too easy, Gillett observes, to become blinded to the "ongoing work of 'self remaking' that someone like Nick is doing." To hear Nick speak through his brother (who has learned to interpret for him), it is obvious that despite severe disabilities, Nick is as mentally quick as ever—and that however we choose to define personhood, he hasn't lost it. Before his accident, he thrived on social interaction and on cracking wicked jokes; if anything, those aspects of his mental life have become more nourishing to him than ever.

"I don't think I could've made it this far without the support of my friends, carers, and family," Nick says. After his return from the hospital, Nick was able to regularly attend his old rugby team's games with the help of a good friend, and his injury did not get in the way of his pulling outrageous stunts. (Nick, his brother, and several buddies landed on You Tube, for example, for showing up naked to the polls on election day.) And it is not just family and old friends who keep Nick going socially. "I have met a lot of people since my accident," he says. "Some have become friends; some have become great friends."

By all accounts, Nick's recovery has been remarkable. During his two years in the hospital, he had to exert an enormous amount of effort to make a single sound, but in 2005 he wrote that "I can now say heaps of words, count, pronounce about four carers' names relatively clearly, sometimes stringing some words together when lying down and relaxed." Since this account in *BMJ*, Nick has learned to use a walker for short distances and even to use certain weight machines with assistance, defying specialists who told him that he would never recover any physical skills. (He remembers one specialist who told him that he'd never move or talk again. When Nick returned home from his long hospital stay, the specialist examined him and was taken aback by Nick's progress. He apologized to Nick for his early predictions. "I gave him the finger," Nick reports.)

His recovery has been extraordinary, but not atypical for LIS patients receiving an intense and consistent level of care. A 2003 study followed 14 patients with locked-in syndrome in three rehabilitation centers for periods ranging from five months to six years, and found that early and intensive rehabilitation treatment significantly improved outcomes and reduced mortality rates.[33] Laureys and colleagues have reported that in a study of 95 patients with LIS,[34] 92 percent recovered some ability to move their heads, and half recovered limited speech production (single intelligible words). Some patients recovered the ability to make small movements in fingers, hands, or arms (65 percent), and three-quarters recovered the ability to make small motions in their legs or feet.

Will to live

Even for someone like Nick, gifted with a fighting spirit and keen sense of humor, there are the inevitable moments when life feels unbearable. Before the accident, reports one of his best friends, he had "muscles on his muscles, he was happy as hell, he had a beautiful girlfriend. He couldn't have been any better."[35] Now Nick says it can be a lonely existence, and he believes that dating and romantic relationships are out of the question. "It would be great," his brother translates, "but able-bodied people struggle getting dates. So I'm stuffed."

He talks of the utter humiliation of his condition—times when he loses control of his bowels in public places, for example. "Believe me, when you're 30 it's totally degrading," he says. "And nowhere more so than in the public gym, in front of people. It definitely changes my mood extremely quickly when it happens, as you can imagine."[36] He admits that he has often thought of suicide, especially alone in his bed at night. Sometimes he wishes that he had died in the ambulance on the way to the hospital—it would have been less frustrating that way—but even if he wanted to commit suicide now, he couldn't do it. "It's physically impossible," he notes—though he has discussed the topic with the people closest to him.

"Yeah, we've talked about it," his brother says. "Nick has said things along the lines of, 'I don't want to live 40 or 50 years in this chair, in this cocoon.' And I can understand that. Who am I to say, in my situation, that he should live his life like that?"[37] But even if it were possible to end his life—alone or with help—Nick says it wouldn't matter. "I'm just glad to still be alive—most of the time anyway," he says. "I only live for hope of recovery now."[38]

Despite the common misperception that LIS patients—given the option—would choose to die, Nick's determination to live fits with the majority sentiment in patients with his condition. "Healthy individuals and medical professionals sometimes assume that the quality of life of an LIS patient is so poor that it is not worth living,"[39] say Laureys and colleagues. "On the contrary, chronic LIS patients typically self-report meaningful quality of life and their demand for euthanasia is surprisingly infrequent." Experts in LIS are concerned that uninformed physicians might provide less aggressive medical treatment than might be warranted with particular patients, despite studies showing decreased mortality and improved quality of life with early diagnosis and treatment. Physicians less familiar with the experiences and wishes of LIS patients might even influence the family toward removal of life support without ensuring that everything reasonable has been done to reveal the patient's preferences.

Regardless of what care locked-in patients ultimately request, Laureys and colleagues stress that their autonomy should be considered paramount. "Patients suffering from LIS should not be denied the right to die—and to die with dignity—but also, and more importantly, they should not be denied the right to live—and to live with dignity and the best possible revalidation, and pain and symptom management."

For his part, Nick doesn't want it left up to anyone else to decide whether his life is worth living. When asked by TV New Zealand in 2007 if he considers himself happy, Nick said, "Absolutely. Coming from where I've been, this is absolutely fantastic. I'm probably happier than most able-bodied people."[40]

Speaking without a voice

Nick's tool to communicate with the outside world is a large transparent board with the letters of the alphabet spread across the surface. He spells out each letter of his sentences painstakingly with his eyes, and the person "listening" on the other side of the board must guess which letter Nick is staring at. This goes on until he has spelled out entire sentences—a process Nick describes as "extremely laborious."[41] For a person like Nick—to whom horsing around with his friends is so important—the inability to banter can be excruciating. It is "very difficult (almost impossible)," he says, "to express yourself or be sarcastic."

Speech devices like Nick's board require assistance from other people to transmit anything verbally—including thoughts and feelings, calls for help, and requests for information—a frustrating reality that only compounds the utter dependence of LIS patients upon others. Cases like Nick's, Gillett says, require that we care enough to "rebuild the tools of communication (through interactive technology and massive personal commitment) so that he can begin to live again among us, albeit with severe disabilities."

State-of-the-art patient-computer interfaces, such as infrared eye sensors coupled to on-screen keyboards, are improving the lives of many LIS patients and their caregivers. These devices allow LIS patients the freedom to perform simple, everyday actions the rest of us take for granted, such as turning lights and appliances on and off, communicating by phone or fax, surfing the Web, sending e-mails, and using word processors and speech synthesizers. These devices cost thousands of dollars—sometimes a prohibitive sum for patients whose insurance won't pay for them—but for those patients who can manage the cost, these devices provide unprecedented freedom.

Patients like Nick—with the ability to consciously control their eye movements—would clearly benefit from the independence such devices afford, but what of patients like the young English woman, who showed signs of consciousness without the ability to control her eyes? For some patients, computer interfaces relying on eye control won't take communication technology far enough. Researchers working with disorders of consciousness welcome the prospect of direct mind-computer interfaces based on fMRI or other functional neuroimaging that could, in effect, read patients' thoughts and emotions, allowing them an active voice in their treatment and in critical life-extending—or life-ending—decisions.

"The beauty of medical and communication-technological progress for patients with severe neurological conditions is that it makes them more and more like the rest of us,"[42] Laureys and fellow researchers with the Coma Science Group observed in a recent editorial. "As a corollary, we caregivers not only owe them the same respect in terms of their autonomy as all other patients, but we also have to rise to so far seldom attained levels of clarity in matters of life and death."

The future of consciousness imaging

The British and Belgian team's conclusion that the young English patient is "beyond any doubt" aware has sparked scientific controversy, but neurologists agree on one point: Her remarkable story strengthens the case for the use of fMRI as a diagnostic tool in the absence of external signs of responsiveness.

The concern remains that with time and additional testing, brain scans might prove unreliable and might even raise false hopes in futile cases. But there is widespread optimism that Dr. Owen's claim might be true—that he and his team "have found a way to show that a patient is aware when existing clinical methods have been unable to provide that information."[43]

In addition to its potential diagnostic uses, Owen, Laureys, and colleagues believe fMRI could prove therapeutically powerful. The young English woman's demonstrable mental actions, they argue, suggest "a method by which some noncommunicative patients, including those diagnosed as vegetative, minimally conscious, or locked in, may be able to use their residual cognitive capabilities to communicate their thoughts to those around them by modulating their own neural activity."[44]

A patient's thoughts could show up on-screen in ways that might be understood by others—a heartening prospect for families and friends of patients like Kate Bainbridge and Nick Chisholm, who may spend excruciating months or years hoping and praying for their loved one's recovery. Even if fMRI fails as a communication device, it might ensure that patients receive the most appropriate care for their condition, as soon after their injury as medically possible.

At the other end of the spectrum are patients who, tragically, will not get better. For these patients, high-tech life support maintains their body without promoting meaningful recovery. The arrival of powerful life-extending technologies in the 1960s and 1970s provided new hope for seriously ill and injured patients who might recover with time, but there was a new problem. These machines could, in the words of Supreme Court Justice Antonin Scalia in the 1990 Nancy Cruzan verdict, "keep the human body alive for longer than any reasonable person would want to inhabit it." Families were faced with the unbearable decision of when, if ever, to withdraw medical care from a loved one.

What parent would not sympathize with the heartache of Nancy Cruzan's father? Nancy—whose car flipped on an icy, deserted road in the winter of 1983 and threw her facedown in a watery ditch—suffered severe brain damage due to oxygen starvation, just as Theresa Schiavo would several years later. Mr. Cruzan, faced with the horrific choice of whether or not to remove his daughter from life support, mourned, "If only the ambulance had arrived five minutes earlier or five minutes later."[45]

Loved ones face precisely the same agonizing dilemma today. The fMRI scan answered a question for the young English patient's loved ones, a question that torments family and friends at the bedsides of comatose and vegetative patients everywhere. "Can she hear what I am saying?"

In this extraordinary case, the answer appears to be "Yes."

2

The good, the bad, and the ugly: powerful emotions and our power to work with them

According to a growing number of researchers who investigate how powerful emotions affect the brain, even if we're awake and "conscious," we might not be acting consciously at all. Anxious or fearful, angry or depressed, or just downright stressed, we can feel more like a slave to our emotions than the boss of ourselves.

Many of us link our emotional tendencies tightly to our sense of personal identity—making matters worse if we happen to be wrestling with especially painful or destructive emotions. We might even feel that we *are* our emotions, that other people are *their* emotions, that we are somehow trapped in a fairly rigid emotional identity, one that's not all sunshine and roses. "He's a hothead," we hear people say. "She's so negative," as if a person's temperament is as much a signature of who she is as her hazel eyes, high cheek bones, or pale complexion. Until recently, conventional scientific wisdom has fallen in step with this notion that our temperament—our "natural" baseline of contentedness or irritability—was set in stone by late adolescence.

Not so, says Richard Davidson, neuroscientist at University of Wisconsin-Madison and pioneer in the field of emotions imaging. "There is tremendous potential for plasticity and for change,"[1] says Davidson, who also sees tremendous potential for this new knowledge about our flexible brains to transform the health care system and—at the early end of the prevention spectrum—our entire educational system as well.

"What I mean by neuroplasticity," says Davidson, "is the fact that the brain is the one organ that is built to change in response to experience. Neuroplasticity is the most important general discovery in all of neuroscience in the last decade. More than your heart, your kidney, your liver, the brain is built to change in response to experience and in response to training. And it is really because of this active neuroplasticity that we can learn."[2]

What makes for this tremendous ability of the brain to change in response to experience? "We now know," says UCLA psychiatrist Daniel Siegel, "that neural firing can lead to changes in neural connections, and experience leads to changes in neural firing."[3] No one denies that our cognitive and emotional makeup are influenced by our genes and early experiences in important ways, but at issue is emerging evidence that our emotional lives need not be dictated by our genetic heritage—nor are all kinds of debilitating emotional damage from our childhoods irreversible. "Emotions are not actually facts," Davidson explains, and by simply being aware of their changing nature, we can de-identify with them, "making it easier to let them go."[4]

Davidson and others in the fields of neuroscience, psychiatry, and psychology have begun demonstrating extraordinary things about the brain's ability to transform its response to emotional stress, thereby lessening physiological stress on the rest of the body. Pictures of the brain in action have shown that connected, loving relationships buffer the way the brain feels fear, and that expectations of manageability can ease the experience of physical pain. They've shown that cognitive therapy can retrain the depressed, angry, fearful mind to be skeptical of destructive thoughts, addressing the age-old question: Is temperament fixed, or can people really change?

Increasingly, the answer appears to be yes, we can—with time, sustained effort, and conscious intention.

Weathering the storm

Haven't we all, at one time or another, felt bullied by negative emotions? Take anger, for starters—maker of wars and lifelong enemies. Studies tell us that being too angry can wreck our physical health and relationships, even end our lives prematurely. A 2004 study found that men prone to feeling hostility or anger were at much higher risk of having a stroke or dying. That study, published in the journal *Circulation*, showed that hostile men had a 10 percent greater risk of developing a heart flutter called atrial fibrillation, a condition that can increase the risk of stroke. For men who unleashed their anger on others, the picture was even bleaker: They were 20 percent more likely to die from any cause during the course of the study.[5]

What about trying to stay healthy on the receiving end of anger? Four children die each day in this country as a result of abuse—three out of four of whom are under age 4—and 80 percent of young adults who have survived abuse suffer from at least one mental health problem (including clinical depression, anxiety or eating disorders, or post-traumatic stress disorder).[6] One out of every four women experiences domestic violence in her lifetime, and nearly one-third of female homicide victims are killed by an intimate partner. Victims of psychological abuse are more likely to experience poor physical health, difficulty concentrating, emotional problems, poor work or school performance, drug or alcohol addiction, and thoughts—or acts—of suicide.[7]

Knowing that chronic anger is lousy for us and for the people we love is one thing; knowing how to tame the beast is another. Even if we're up on the research about aggression's harmful effects, sometimes we just can't help ourselves. We snap at our partners and our kids, we curse the driver who steals our parking spot, we seethe when a coworker takes credit for our ideas, or we feel righteous anger at the political candidate we detest the most. Anger might be one of the most destructive emotions around, but when we're caught up in it, it can feel invigorating. It gets the juices flowing.

As it turns out, there might be a healthy reason for the instinct to go ahead and feel a strong emotion of any kind—just long enough to start working with it. New research shows that when we're caught in the grip of a powerful emotion, simply repressing the desire to express it isn't great for us either. Expressive suppression is defined as the conscious effort to inhibit the overt expression of emotions, and recent research on this mode of emotion regulation supports the long-standing psychological wisdom that pretending to be okay is never a fabulous emotional strategy.

"Road rage, office rage, and even air rage are now regularly in the news," says Stanford psychologist James Gross, "providing compelling anecdotal evidence of the damage that is done by failures to regulate negative emotions such as anger. The grim statistics on spousal, child, and elder abuse stand as a further testament to the serious harm that can come from dysregulated emotions."[8] Acting impulsively has obvious ill effects, but Dr. Gross and colleagues have shown that suppression is not a healthy way to go either. It impairs memory function and fails to decrease the internal experience of a negative emotion, yet it manages to decrease the experience of positive ones.[9] It also elevates blood pressure and other physiological stress markers—not just for the person suppressing his emotional response, but for his intimate or social partner as well.[10]

Where does all this new information leave those of us a little too quick to the trigger? If yelling at an irritating neighbor is a terrible idea—but so is pretending we don't want to—whatever is a hot head to do?

A growing number of researchers are looking somewhere between the two extremes of reactivity and repression, hoping to fill our emotional toolboxes—first by using new imaging technologies to figure out what's happening to our brains as they weather intense emotional storms, and then by watching them react to different attempts to steady the ship. How do sad brains react to antidepressants? How do they react to cognitive therapy? How do panicked brains react to beta-blockers, traumatized brains to mindfulness practice, angry brains to forgiveness practice? These are some of the questions researchers are asking about our emotions, and images of the brain are providing some intriguing answers.

Watching our minds watch themselves

At first glance, the concept of mindfulness—the simple practice of watching our moment-to-moment experiences, including thoughts, emotions, and bodily sensations—might seem too new-agey, too pacifist a strategy for someone grappling with intense emotions, but take another look: Recent imaging research is showing that this rather quiet, subtle practice might just turn out to be one of the mightiest tools to change the brain. Akin to the concept of metacognition in

Western psychology (defined as awareness of thoughts and of thinking processes), mindfulness is sometimes described in Buddhist psychology as "bare, nonjudgmental awareness," and it is the subject of a revolutionary new functional MRI study performed at the University of Toronto.

Researchers wondered if people with mindfulness training would deal with sadness differently than people without experience with the technique, and if so, how exactly would those differences appear in the brain? "We induced sad moods in people while they were in the scanner, and we looked at how they regulated themselves in the face of these sad feelings,"[11] recounts psychologist Zindel Segal, one of the study's authors. Participants came from two groups: One group had recently completed a course in mindfulness-based stress reduction (MBSR), and another was waiting to take an identical course. All participants were patients undergoing treatment at a Toronto clinic for depression, anxiety, or chronic pain.

MBSR is an eight-week treatment program for chronic stress, pain, and illness developed by Jon Kabat-Zinn, prominent American author, scientist, and meditation teacher and founder of the Stress Reduction Clinic and the Center for Mindfulness in Medicine, Health Care, and Society at the University of Massachusetts Medical School. The program features meditation and yoga practices that calm body and mind and build the capacity to attend to the present moment with nonjudgmental awareness.

Starting in the 1990s, Dr. Segal began working with Dr. Kabat-Zinn and two colleagues at Oxford University, psychologist Mark Williams and psychiatrist John Teasdale, to integrate MBSR techniques with cognitive therapy for the treatment of mental health conditions. They dubbed the collection of techniques mindfulness-based cognitive therapy (MBCT) and set out to study its effectiveness in the treatment of clinical depression. In a large study of patients who had recovered from two or more episodes of major depression, 65 percent of those who received mindfulness training remained stable after a year compared with 34 percent in a control group. In patients with three or more episodes of depression, the relapse rate was halved.[12]

Why, exactly, were mindfulness techniques better at preventing relapse? This was still speculative territory. When Segal and colleagues at Toronto designed the fMRI study, they hoped to shed new light on the real-time effects of mindfulness practice on brains grappling with sadness. The pictures did not disappoint.

"What we found," Dr. Segal said, "was that if people were feeling sad, then there were some parts of the brain that automatically came online to try to help figure out what was going on."[13] These tended to be areas involved in self-reference, planning, threat evaluation, and problem solving—areas, Segal said, which might tend to questions like, 'Is this a problem for me? Do I need to do something about this? What does this mean about me?' etcetera. That happened to everyone at the outset." Whether or not patients had undergone mindfulness training, sad was sad, distressed was distressed. Everybody felt it.

An interesting finding in and of itself, but then things got fascinating: During the course of their time in the scanner, the mindfulness group dealt quite differently with that initial emotional response than did the control group. They were able to reduce the intensity of activation in strategizing, self-ruminating areas, and increase information coming from bodily sensations. "It was like they were opening up a second channel, so the experience of emotion was not one that was just mediated by these kinds of verbal, strategic areas—'What do I do about this?'"[14] There was still a reduced level of that kind of activation, Segal said, but other areas of the brain pitched in as well, areas that tended to questions like, "'How does the body feel at this point? What else is going on in my awareness of the body, posture, position, breath?'" The group that had been introduced to mindfulness practice, Segal suspects, had access to "a more complete picture of what was happening from moment to moment," and less mental energy was diverted toward rumination or self-focused thought.

Self-referential thought has been implicated in prolonged distress after an emotionally challenging experience, as compared to strategies that reduce obsessive thinking about, or elaboration upon, a negative experience.[15] Reducing self-referential thought is at the heart of traditional Buddhist approaches to treating emotional illness, and the Toronto fMRI findings point to some strong reasons for this strategy's effectiveness.

"That seems to be one of the things that mindfulness, as we understand it now, is helping people do," Segal says. "It doesn't shut off the more automatic activations that come from dysphoric moods, but provides an augmented or enhanced image, moment-to-moment, of what actually is occurring."[16]

Dr. Mark Williams, professor of psychiatry at the University of Oxford and co-creator of the MBCT program, says that mindfulness is such a powerful technique because it "teaches a way of looking at problems, observing them clearly but not necessarily trying to fix them or solve them. It suggests to people that they begin to see all their thoughts as just thoughts, whether they are positive, negative, or neutral."

But can simply *watching* a powerful emotion act on the mind and body really lessen its impact? Williams thinks so. Mindfulness, he says, "involves dealing with expectations, with constantly judging ourselves—feeling we're not good enough.... All of these things are just thoughts. And, they will come up in meditation and learning to recognize what they are as thoughts, and let them go, can be enormously empowering for anybody."[17]

Retraining the traumatized mind

If MBCT can retrain a depressed mind, what about a panicked mind caught in the grip of a traumatic memory? Post-traumatic stress disorder (PTSD) is a debilitating anxiety disorder that can develop after a terrifying incident involving physical harm or the threat of physical harm—either to the person who develops PTSD or to anyone the PTSD sufferer witnesses being harmed. Many people with PTSD relive the traumatic event in their memories during the day and in nightmares when they sleep, and a person caught in the grip of a "flashback" might feel as though the traumatic incident is happening all over again.

PTSD affects about 7.7 million American adults, but it can occur at any age, including throughout childhood, and it is often accompanied by depression, substance abuse, or one or more other anxiety disorders. PTSD can develop in response to a variety of traumatic events. The National Institute of Mental Health gives the following examples of precipitating causes: "Mugging, rape, torture, being kidnapped or held captive, child abuse, car accidents, train wrecks, plane crashes, bombings, or natural disasters such as floods or earthquakes."[18]

Although PTSD can result from a number of terrible events, the condition has received a great deal of attention of late due to its prevalence among veterans returning from the Iraq War. A 2004 study published in the *New England Journal of Medicine* found a "strong reported relation between combat experiences, such as being shot at, handling dead bodies, knowing someone who was killed, or killing enemy combatants, and the prevalence of PTSD."[19] Among soldiers and marines in the Iraq War, the prevalence of PTSD when they returned home increased linearly according to the number of firefights they experienced during their deployment: 4.5 percent of soldiers experienced PTSD even with no firefights; 9.3 percent did for one to two firefights; 12.7 percent did for three to five firefights; and 19.3 percent did for more than five firefights. These are chilling numbers, and increasingly, brain researchers are looking to the neural patterns of veterans and others suffering from the debilitating effects of trauma in order to understand what has been damaged and how best to fix it.

One lab at the University of Michigan, headed by psychiatrist and neuroscientist Israel Liberzon, is investigating how the brains of veterans with PTSD respond differently to reminders of their trauma than the brains of people without PTSD. His group has looked at a range of questions about trauma and how it affects the brain: How does it shape the experience and expression of emotion, for example? How does it alter stress hormone systems? How do traumatic events in childhood contribute to vulnerability to psychiatric disorders in adulthood, and do they affect people's ability to parent? "We're interested in what happens when bad things happen to good people,"[20] said Anthony King, neurobiologist with the group who is analyzing whether mindfulness-based therapy might be an effective treatment for PTSD.

To get at the neural underpinnings of the condition, Liberzon, King, and colleagues designed a study to watch how the brains of combat veterans with the condition might respond differently than people without PTSD to reminders of their own trauma and to distressing images unrelated to their personal stories. They wondered, for starters, whether specific parts of the brain were in charge of down-regulating the neural stress response, and whether these areas were sufficiently engaged in people with PTSD. "Parts of the brain might be a gas pedal or accelerator, which could trigger the stress response," said King. "Other parts of the brain might be inhibitory, like a brake."

The researchers performed PET scans to see how potentially stressful triggers affected three different study groups: veterans who had served in combat and suffered from PTSD, veterans with comparable combat service who did not suffer from PTSD, and veterans who had never served in combat and did not suffer from the condition. Subjects listened to recordings they had made recounting traumatic events from their lives, and at separate intervals, they also viewed gruesome pictures unconnected to their personal narratives. "These are some gory pictures," King observed, which conceivably could trigger memories from bloody combat experiences. But they are "not personally relevant," he said, "and they don't involve a sense of self very often."

The gory pictures, as it turns out, had no observable effect on stress hormone responses, but the personal trauma narratives did. This was true for both groups of combat veterans—indicating that the magnitude of the stress hormone response was not a main effect of PTSD, but instead was simply due to remembering severe personal trauma with or without the presence of PTSD.

The brain images, however, revealed striking differences between the two combat groups: The veterans without PTSD showed a much stronger coupling of stress hormone levels with the medial prefrontal cortex (the "third-eye" area right behind the forehead), whereas the PTSD patients showed a much stronger relationship between hormone levels and an area of the brain called the insula—part of the limbic system believed to be a key interface between physical sensations and emotions—as well as a relative deficit in medial prefrontal cortex activity.[21] These results, King said, "were suggestive of the idea that there's a deficit in regulation [in PTSD], that there's not a good braking system for emotional responses."

In light of emerging evidence that mindfulness practices alleviate symptoms of chronic pain and anxiety and that mindfulness-based cognitive therapy makes depressive relapse less likely, King was curious to see if mindfulness training might prove similarly powerful in treating PTSD symptoms. Experiencing intrusive thoughts and feeling completely overwhelmed by them are classic symptoms of PTSD. These painful effects can happen at any time or place, in response to a memory or to some other reminder of the traumatic event—the evening news, for example, or any other evocative sensory experience. "A lot of Vietnam vets are very strongly triggered by hearing helicopters, or by smelling diesel fuel, or sadly, even by hot weather," said King, pointing out that many Vietnam veterans have an extraordinarily difficult time in the summer.

King, an experienced meditation practitioner in the Tibetan and Korean Zen traditions, wondered if certain mindfulness practices—watching painful memories arise, for example, and watching our emotional and physical reactions to them—might alleviate the otherwise devastating effects of memories that are intimately linked with excruciating emotions. "There's a core practice of mindfulness of just letting things be and noticing that they don't last forever," he said. "They actually do pass. The upsetting thought eventually does go away. This can be an eye-opener for a lot of people."

To test the efficacy of mindfulness training for PTSD, he ran a pilot study of eight-week group courses that introduced veterans to several mindfulness techniques, including the "three-minute breathing space," a technique first presented by Williams, Teasdale, Segal, and Kabat-Zinn as part of their MBCT course for depression. This breathing exercise can be done at brief intervals throughout the day to "give yourself a little attitude adjustment," as King described it, and he said that as a group, the veterans found it extremely useful. He also introduced the practice of *metta*, which is the practice of consciously cultivating unconditional love for oneself and others. Usually translated into English as "loving-kindness," this practice involves the contemplation of phrases like, "May I be happy," and "May my children feel safe and protected," and so on, focusing attention on self and loved ones, friends, neutral and difficult people, and eventually extending the practice to all beings.

"I was really kind of going out on a limb," King said, and he wondered, "What are these old combat vets going to do? Are they going to walk out, or what's going to happen? Turned out all of them really appreciated this." The whole idea, he said, was to cultivate positive emotions for people who often experience severe emotional numbing as a result of their condition—"almost an inability to feel things like love or happiness." He asked them to begin by sending metta to someone beloved, someone who made them smile just to think about. "They really got that. It was really a positive experience for most."

Sending metta to themselves, however, was a different story. The fact that it was a challenging practice for many participants was unsurprising because PTSD often has a substantial amount of guilt and self-blame associated with it. "One of the biggest cognitions that predicts who's going to get PTSD is thinking it's my fault," King said. "'This happened because of the type of person I am,' whether it's sexual assault or whether it's combat. 'Bad things happen to me because there's something wrong with me.' Obviously that's a pretty distressing type of thought to have."

Despite some practices the veterans found difficult, at the end of the pilot period, King and fellow researchers found a significant reduction in PTSD symptoms after the eight-week group courses. This was especially encouraging, King said, because published data on other types of group therapy for PTSD has not shown a significant impact on symptoms— although groups are beneficial for other things, such as building a sense of community and the reassurance that comes from being in a group of people who have shared similar experiences. "Coming up with an effective group would be a really good thing," King observed.

What's more, almost all the improvements were seen in what are called avoidance symptoms. Avoidance is one of the most powerful coping strategies PTSD sufferers develop, and they often become quite skilled at evading thoughts and emotions by staying busy all the time, by never allowing themselves the opportunity to just sit and reflect. For traumatized vets, the space and time to think might mean encountering painful memories and the gripping emotions they ignite. "Many of them have built really successful coping strategies by keeping themselves absolutely busy every second of the day, and it usually works up to a point," King said. "They tend to work very hard, do a lot of overtime, work 15, 16 hours a day. And when they're younger, they drink to get to sleep. What happens is at some point, their body gives out." With mindfulness practice, however, participants learned ways to stay present with upsetting memories and emotions rather than simply avoiding the experiences.

King and colleagues have now embarked on a more extensive, four-year study in consultation with Kabat-Zinn, Segal, and others to test if mindfulness-based therapy is as effective for these symptoms as the pilot study suggests. The new phase will include a pre- and post-therapy neuroimaging component, much like the imaging studies that have been performed for cognitive behavioral therapy for depression.

Why go it alone?

If the ability to self-regulate strong negative emotions is vital to our health and well-being, so too is our ability to promote strong, support-ive relationships. Public health researchers have long identified social isolation as a serious health risk, and among married people, higher marital quality correlates with decreased risk of infection and faster recovery from injury—even decreased death rates from life-threatening illness. Why is this so?

"Neuroplasticity enables us to be influenced by and changed by the experiences we have, by the people around us," Dr. Davidson says. "Wittingly or unwittingly, we are always changing each others' brains."[22] Supportive social behaviors have been shown to alleviate physiological distress and negative emotions, and Davidson, along with James Coan of University of Virginia and Hillary Schaefer of University of Wisconsin-Madison, set out to show how strong rela-tionships and supportive behavior can buffer the experience of fear.

The study worked like this: Sixteen happily married women were subjected to the threat of electric shock while holding their husband's hand, the hand of a male stranger, or no hand at all. Functional MRI scans of their brains were recorded in each case, and the results were clear. In the face of a perceived threat, hand-holding alleviated the women's anticipatory stress in areas of the brain central to the neural fear response. And the better the marital relationship, the weaker the neural fear response. Hand-holding with a stranger was better than no hand-holding at all, according to measures of bodily arousal, although only holding the hand of a spouse lowered the women's reported experience of "unpleasantness." Holding a stranger's hand appeared to decrease the need to produce a coordinated bodily response to threat—in other words, the basic instinct to get out of Dodge. Hand-holding with a spouse, however, was "particularly powerful" in that it conferred "the additional benefit of decreasing the need for vigilance, evaluation, and self-regulation"[23] of negative emotion.

Healing brains, healing institutions

The take-home message of this encouraging research is that many of the damaging physical and mental effects of our most powerful negative emotions—fear, anxiety, sadness, and anger—are not preordained, nor are they irreversible. This is great news, especially for those of us more prone to strong negative emotions, but the question remains: How will this new information about our flexible brains be incorporated into major cultural institutions, like health care and education?

"I think that it is going to be a series of evolutionary changes," Davidson says. "I think it's going to take a while."[24] Our standard model of psychological treatment, he feels, "is really completely inadequate. And I think that it's inadequate because of the failure to adequately account for the role of practice in producing neuroplastic changes. Some changes in the brain are changes that will only come about with extremely systematic, rigorous, serious practice."

Our standard treatment model—once a week for 45 minutes—is "just completely ridiculous, based on everything we know about the brain," Davidson says. "People know that if they do physical exercise—physical exercise once a week is just not going to cut it. And so why should this be any different? That's a message that I think needs to be just hammered away at."

It is a message that Davidson thinks people are starting to connect with, and not a moment too soon. "I do think we would see happier, better adjusted kids if these practices were more widespread and incorporated into our major institutions of the culture." There are encouraging pilot studies of contemplative education happening across the country, including in some of the largest school districts. "There are 27,000 New York City school children who are currently being taught contemplative practices," Davidson notes, "and these are all kids who were personally affected by 9/11."

The standard model of treatment might be slow to change, but a growing number of individual therapists are finding ways to integrate contemplative practices like mindfulness into treatment of psychiatric disorders. "The interest in this has just taken off," Segal says. "And I think a big part of it is that more and more therapists are practicing some form of contemplation themselves and want to bring that into therapy."[25]

Aliza Weinrib, a clinical intern on staff with Dr. Segal, practiced meditation for years before becoming a therapist. "From my personal experience," she says, "I believed that mindfulness and meditation practices could improve mental health and quality of life." Her own contemplative practice "created a kind of space between me and my 'negative' emotions. They were there and I could feel them, of course, but they did not control me in the same way. I had more freedom to act. I could notice my distress, and still do what needed doing. Somehow, over time, this reduced my distress as well."[26]

These psychological benefits, she believed, would probably translate to clinical practice in some natural ways, but mindfulness was not a hot topic of conversation in her graduate seminars. Then she happened upon a class with a psychotherapy instructor who was interested in how mindfulness techniques can improve the therapist-patient relationship. "I had been studying with him for two years and had no idea he was interested in [mindfulness]; he had been teaching me for two years and had no idea that I was practicing it." She was already employing mindfulness techniques in her work—not as an explicit therapeutic tool, but as part of her preparation for a session. "I found it very helpful in coming to a state of open attention to the person I was working with—but I didn't talk about this with anyone. Not because it was a secret, but because I didn't think it was something that anyone else would be familiar with or interested in." The seminar exposed her to a growing community of therapists who were investigating clinical applications of mindfulness, and she no longer felt isolated in her interests. "That is when I started to incorporate mindfulness into my work as a therapist and a researcher in earnest. Encountering psychologists who were also mindfulness practitioners made me a better psychologist and, to my surprise, also deepened my mindfulness practice."

The recent fascination with contemplative practices is not limited to the treatment of mental and emotional health. Staff at the Mayo Clinic met with the Dalai Lama, Richard Davidson, Jon Kabat-Zinn, and others in the spring of 2008 to discuss the use of mindfulness and other contemplative practices in medical treatment and prevention. The staff was "fantastically receptive," Davidson says. "They understand that this is something that is going to complement traditional medical practice and will help with patient care and will help with the bottom line. There are data indicating that [contemplative] practices actually speed wound-healing.... They will likely decrease health care costs. There is every reason to believe that would be the case."[27]

As of summer 2009, NIH is funding more than 60 studies of mindfulness techniques—up from three in the year 2000—to treat health conditions ranging from depression to addiction, asthma to irritable bowel syndrome, heart disease to hot flashes.

3

Happiness on the brain

Happiness makes us—happy. It's our favorite state of mind. We might not be experts at finding it, but we know it when we see it—and we'd like a map to the rest of it, please. (We promise to share. Happy people are generous, too.)

The brain's power to heal from illness and injury couldn't be a hotter topic these days, but what does this uplifting insight into our neural resiliency mean for those of us walking around with a "normal" baseline of mental health? Even if we've never battled chronic anger or anxiety, depression or panic—thanks perhaps to cool temperament, smooth circumstances, or a lucky combination of both—even the steadiest among us might wonder if there's more to life than just riding the waves. Is there such a thing as better than okay?

Martin Seligman, former president of the American Psychological Association (APA), wondered, too. "I realized that my profession was half-baked. It wasn't enough for us to nullify disabling conditions and get to zero. We needed to ask, 'What are the enabling conditions that make human beings flourish? How do we get from zero to plus five?'"[1] In the late 1990s, Seligman was instrumental in the rise of a new school of thought about the power of thought—about just how healthy our minds could get under ideal conditions, and what, exactly, those conditions might be.

His inaugural act as APA president was to invite Ray Fowler, CEO of the APA, and psychologist Mihaly Csikszentmihalyi, a co-conspirator in happiness research, to a Mexican beach to talk about positive psychology and the future of the field. They left the meeting with plans for the first-ever conference on positive psychology to be held the following year (on the same gorgeous Mexican beach—go figure). With a new call to service from APA leaders, researchers throughout the country set out to explore the building blocks of well-being, their collective curiosity triggering an explosion of research into which patterns of thinking, types of relationships, and ways of behaving in the world are conducive not only to temporary pleasure, but to lasting happiness.

Seligman's own work on optimism had convinced him that there is not just one major component of happiness but three. There's pleasure, of course—as basic as the delight we experience when the first piece of chocolate cake graces our tongue, or as heady as the pride we experience when our boss praises our work. Pleasure is qualitatively different, Seligman argues, than the happiness we derive from deep engagement with our family, our work, and our other passions, and it is also distinguishable from the happiness we derive from "meaning"—which he describes as the satisfaction of applying one's strengths and talents for a higher purpose. "This is newsworthy," Seligman says, "because so many Americans build their lives around pursuing pleasure. It turns out that engagement and meaning are much more important."[2]

Other researchers have drawn different distinctions—between, for example, hedonic well-being (life satisfaction and sunny outlook) and eudemonic well-being (a sense of purpose, growth, and mastery, which need not be accompanied by that warm fuzzy feeling). However you slice the happiness cake, if there are qualitatively different sources of well-being, shouldn't they look different when we image the brain?

When positive psychology was emerging as a new academic field, Richard Davidson had already devoted nearly a decade to looking at how different emotional states work within the brain. He welcomed the new elbow room for positive mind states. "In their recent admonition to the field," Davidson and colleagues wrote in 2004, "Seligman and Csikszentmihalyi exhorted researchers to direct efforts toward understanding the processes that enable humans to flourish, even under benign conditions. Indeed, that which makes a life worth living surely encompasses those aspects of the human condition that denote happiness, fulfillment, and enrichment—well-being."[3]

Prior emphasis on negative emotions, Davidson noted, was based in part on the faulty assumption that a mere absence of negative mental processes implies a healthy, "adaptive" mental life. Thanks to new images of positive emotions, we can see that the ability to regulate negative mental states is only half the picture. Emotions such as love, empathy, compassion, and joy for others' good fortune—emotions that correlate closely with the subjective experience of happiness— light up telltale networks in the brain. And much like their negative counterparts—anger, sadness, fear, and jealousy—the cultivation of positive emotions can stretch and strengthen key neural networks. "I've been talking about happiness not as a trait," says Davidson, "but as a skill, like tennis. If you want to be a good tennis player, you can't just pick up a racket—you have to practice."[4]

Just as we train our bodies to run for the ball or drill it past our opponent, we can train our minds to be happier in response to whatever challenges—or delights—we might encounter.

The asymmetrical brain

Throughout the 1990s, Davidson's work demonstrated a convincing link between activity in the left prefrontal cortex (PFC) and positive emotional states. Positive affect—or what our parents might have termed a "good attitude"—and emotional resilience also correlated with quick recovery from emotional challenges and the ability to consciously regulate activity in the amygdala (an emotion-processing hub in the brain, believed to be a key player in conditioned learning and formation of long-term memories). And there was more good news for people living on the sunny side: Research subjects with higher left PFC activation showed enhanced immune function when compared to their right-brained counterparts.

This asymmetry in emotional tasking results from the basic biology of approach/withdrawal behavior, Davidson explains. "You can think of approach and withdrawal behavior as capturing the fundamental psychological decision that an organism makes with respect to its environment,"[5] he says. If an animal likes what it sees, it can choose to approach a source of stimulation; if it doesn't, it can take a pass. This night-and-day behavioral choice is especially clear in simple organisms, some of which don't do anything but approach and withdraw.

Behavior gets increasingly tricky when backbones and divided nervous systems call the shots. In 2007, Italian researchers made a fascinating discovery about the brains of some of our closest and furriest friends. Dogs—not just humans and other primates—show a left brain bias when they're feeling especially happy. Upon seeing their beloved owners, their tails wag more to the right side of their rumps (indicating more left brain activation). When they see a strange, aggressive dog, by contrast, their wagging is heavily biased to the left (indicating that the right brain has taken charge in a threatening situation).[6]

With control of complex behavior localized in different hemispheres, animals are able to avoid duplication and maximize efficient use of brain tissue. They can juggle two important actions at once, such as foraging for food (an "approach" or goal-oriented behavior) while being poised to run from predators (a "withdraw" behavior). Approach behavior generally requires much more fine manual control—picking up food or something else of interest, for example. The left hemisphere is the known seat of fine manual control.

For humans and other primates, approach behavior can appear as pointing, touching, grasping, or caressing, and this outward, goal-directed behavior tends to be associated with positive feelings like desire, attachment, and love. It is also associated with physiological markers of security, tranquility, and relaxation, such as slow respiration and heart rate. Withdrawal behavior, on the other hand, is typically a choice to flee or—more subtly—to distance oneself. It tends to be associated with negative feelings like fear and anxiety and with physiological markers of distress, such as rapid heart rate, depressed immune function, and sluggish digestion.

It makes a good deal of evolutionary sense that feeling happy, or fighting off a cold, or nourishing our bodies would take a back seat if we suddenly find ourselves in the presence of a lion or a mugger, but our mother-in-law—not so much. Maybe over Thanksgiving dinner, we'd like our positive emotions to do the driving.

Do we get to decide? How much "give" is there in our happiness muscle, and—if there is room for improvement—how can we bend it in the right direction?

Pleasure is nice, until it isn't

Since the accidental discovery in the 1950s of a "pleasure center" in rats' brains, scientists have known that specific neural circuits are key to the pleasure response. Half a century ago, scientists James Olds and Peter Milner were attempting to identify areas of a rat's brain that would cause discomfort when stimulated with electrodes, but as it turned out, their research produced the opposite result: It identified a spot in the brain that, when shocked, actually brought the rats back for more. The little guys couldn't get enough. In fact, when rats were allowed to push a lever that delivered a shock to this area of the brain—called the nucleus accumbens—they stopped eating and drinking and just pressed that little lever with all their might, until eventually they died of exhaustion.

Recent neuroimaging research has expanded and refined our understanding of the circuitry of reward and pleasure, and it has produced some eye-opening clues to how things can go wrong when the brain experiences too little pleasure in response to pleasant stimuli—or, on the flip side—too much pleasure in response to stimuli that should not feel pleasurable at all.

Anatomy of a milkshake

Who hasn't, at one time or other, caved into the unhealthy desire to finish off an entire bag of chips, or swing through the late-night drive-through, even when we know we'll regret it? One little sip of that midnight milkshake and the brain's pleasure response lights up like the Rockefeller Center Christmas tree, and we like it. We come back for more. Eventually we stop—maybe we get sick and swear off milkshakes for a week. But what happens if we become helpless slaves to those cravings, like little rats pressing a lever?

Obesity is the second leading cause of preventable death in the United States after smoking, and more than a third of American adults are obese. Between 1980 and 2000, obesity rates more than doubled among American adults—and more than tripled among adolescents. Unhealthy trends in diet and exercise are major contributing factors to the epidemic, of course, but obesity also runs in some families. Researchers have long wondered about the role of genetics in overeating and obesity, and malfunctioning reward circuitry in the brain has been a prime suspect. Perhaps people who overeat experience a stronger pleasure response to food than other people, or perhaps their brains get addicted to the pleasure response in a way that the brains of people who maintain a healthy weight do not.

What is really going on in the brain of an overeater, and how might that information be used to help those of us who are vulnerable? Dr. Eric Stice of the Oregon Research Institute used fMRI technology to investigate the problem, and his experiment produced some fascinating—and counterintuitive—answers. A group of young women, ranging from very thin to very obese, tasted milkshakes inside a scanner, and the pictures showed that a key pleasure center deep in the brain, the dorsal striatum, was much less active in overweight people than in lean people and in people who carry a particular gene variant. What's more, the young women who carried the gene—known as Taq1A1—gained more weight in the year ahead than women without the gene.

"The more blunted your response to the milkshake taste," Stice concluded, "the more likely you are to gain weight."[7] Stice and his coauthors theorized that people who experience less pleasure in response to food might overeat to compensate, whereas people who experience more pleasure stop because they are satisfied. Prior studies demonstrated that obese people had fewer dopamine receptors in the brain, but this was the first brain-scanning study to link a suspected genetic factor to observable patterns of brain function.

Dr. Nora Volkow, director of the National Institute on Drug Abuse, has spent years studying the role of dopamine in motivation, desire, and pleasure. She called the study "elegant" because "it takes the gene associated with greater vulnerability for obesity and asks the question why. What is it doing to the way the brain is functioning that would make a person more vulnerable to compulsively eat food and become obese?"[8]

Volkow wonders, though, whether obese people are really overeating to try to boost an underactive pleasure response, or whether a suboptimal dopamine response causes them to be more impulsive than other people. "Dopamine is not just about pleasure," she says, noting that it also plays critical roles in conditioned learning and impulse control.

The brain of a bully

What makes a kid want to be cruel? Any tormented child knows that there is something going on inside a bully's head that isn't right, but what exactly has gone wrong, and what, if anything, can be done about it?

Researchers at the University of Chicago have used fMRI to probe the minds of adolescents diagnosed with conduct disorder (CD)—a pattern of unusually aggressive or violent behavior—and the results were startling. Their work, published in the journal *Biological Psychology*, scanned the brains of eight adolescent boys with CD and a matched control group of eight teen boys without CD symptoms. The boys watched video clips of people causing others pain, sometimes accidentally (a hammer dropped on a person's toe, for example) and sometimes intentionally (a piano lid slammed shut on a player's fingers).

"Aggressive adolescents showed a specific and very strong activation of the amygdala and ventral striatum (an area that responds to feeling rewarded) when watching pain inflicted on others, which suggested that they enjoyed watching pain,"[9] said Jean Decety, lead investigator on the study. More studies are needed, the authors cautioned, before anyone can state with certainty that the boys' heightened neural response was more akin to pleasure than to a negative emotion or sensation.

An alternate hypothesis is that intense activation of the pain matrix in teens with CD causes higher levels of distress, which might in turn lead to ramped-up aggression. Physical pain can elicit aggression, the authors note, and stronger activation of the neural networks that underpin pain processing—coupled with a reduced capacity to regulate negative emotions—might result in aggressive behavior. "For example, youth with CD who see an injured friend (or fellow member of a gang) may be more likely to respond aggressively than other youth for this reason."[10]

Although press reports sensationalized the potential role of the pleasure response, the Chicago investigators' surprise at the results was primarily due to the fact that they had expected to record a blunted emotional response to watching others in pain—an indifference that might allow kids with CD to intimidate or harm others without being hindered by emotions like guilt. "The prevailing view is these kids are cold and unemotional in their aggression," says Benjamin Lahey, University of Chicago psychologist who worked on the study. "This is looking like maybe they care very much."[11] Instead of a neural deficit in response to others' pain, teens with CD actually displayed an "enhanced response to images of people in pain."[12]

Coupled with this heightened response in the pain centers, however, was a clear deficit in the ability to regulate the intensity of their response. When teens without CD saw others in pain, there was a spike in activity in the pain matrix and in the medial prefrontal cortex—a part of the brain key to self-regulation and emotion management. "But in kids with conduct disorder," said Lahey, "that connection isn't there."[13]

Images of the normal teens' brains meshed with earlier work showing that younger kids, ages 7 to 12, feel naturally empathetic toward others in pain, and that they experience empathy as increased activation in parts of the brain that engage when they experience pain themselves. When the younger group viewed images of intentionally inflicted pain, areas of the brain involved in social and moral evaluation came online, and connections to attention networks in the prefrontal cortex increased.

"The full-blown capacity of human empathy is more sophisticated than the mere simulation" of another's emotional state, wrote Decety and colleagues. "Indeed, empathy is about both sharing *and* understanding the emotional state of others in relation to oneself."[14]

Happily ever after?

Even if our pleasure-seeking nature doesn't get us into serious trouble, let's face it: The experience of pleasure can be a rather unreliable source of happiness. We enjoy the feeling of sun on our skin, but we happen to work in a windowless office in Seattle. Maybe we're the type that craves positive reinforcement, but we live with a spouse who missed the "How to Compliment Your Partner and Sound Like You

Mean It" day in relationship school. (If they didn't miss it, we're sure they failed it.)

What a relief, then, that other experiences lead to happiness—like the joy we feel when we connect emotionally with that maddening but lovable spouse, or when we play with our maddening but lovable children, or when we change a friend's or a stranger's life for the better. In a recent poll, people were asked to identify their major sources of happiness, and landing in the top four spots were relationships with children, friends and friendships, contributing to the lives of others, and relationships with partners.[15] Although a quarter of respondents said they'd eat to improve their mood, more than half said they'd choose to talk to friends or family. Do meaningful connections with others make us feel better, at least in part, because they affect our brains differently than momentary pleasures?

Richard Davidson took compassion—a rather neglected emotion in the psychological literature—as his starting point. "If you look at the index of any scientific textbook, you won't find the word compassion," Davidson says. "But it is as worthy a topic of examination as all the negative emotions—fear, anxiety, sadness, anger, disgust—that have long occupied the scientific community."[16]

Davidson has drawn inspiration from the life and work of the fourteenth Dalai Lama, beloved leader of Tibetan Buddhists and international icon of love and forgiveness, who emphasizes compassion above other emotions as an antidote to suffering and aggression in the world. "If you want others to be happy," the Dalai Lama has famously said, "practice compassion. If you want to be happy, practice compassion."[17] Buddhism is not alone among the world's spiritual traditions in its emphasis on this prosocial emotion; most contemplative practices count the wish to alleviate others' suffering among the highest of virtues.

If compassion is so special, Davidson hypothesized, it should do special things to the brain. "We wanted to see," he says, "how this voluntary generation of compassion affects the brain systems involved in empathy."[18] At a meeting in Dharamsala, India in 1992, Davidson forged a partnership with the Dalai Lama that would transform the study of emotion, attention, and well-being in Western neuroscience.

Davidson says it was at that meeting that he "made a commitment to him, as well as to myself, that I was going to now come out of the closet, so to speak, with my interests in meditation."[19]

The high Tibetan leader encouraged seasoned monks to volunteer their brains for study, and beginning in 2000, eight monks—each with tens of thousands of hours of meditation experience—took turns traveling to Davidson's lab in Madison, Wisconsin. They donned elaborate headgear—a sort-of chain mail of highly sensitive electrodes—that measured their brainwaves while they consciously generated states of compassion. (The Dalai Lama, when asked why he had not submitted his own brain for study, reported that his brain would look like "nothing special" because he never has "sufficient time to meditate."[20])

This first wave of experiments landed meditation in the national spotlight and put an intense form of mental training on the map of legitimate neuroscientific inquiry. When experienced practitioners actively generated a state of "nonreferential compassion"—a compassionate state of mind not directed toward any particular being or group—their brains produced high-frequency (gamma) activity like nothing ever measured, and this remarkable activity remained higher in their post-meditation resting state.[21] Gamma activity is believed to be an indicator of neural synchrony—the integration of localized neural processes into higher-order cognitive and emotional functions. "That for us is an indication," says Davidson's partner on the project, Antoine Lutz, "that through the training something happened in the brain. Something had changed in such a way that they can generate these very, very integrative and coherent states during meditation."[22] What's more, the level of gamma activity correlated with accumulated hours of practice, bolstering the hypothesis that the results were, at least in major part, due to the practice itself, and could not be written off as the anomalous work of eight brains that might have started out extraordinary in the first place.

The Madison team had succeeded in proving that the monks' mental engines were uncommonly powerful; the next step was to peer under the hood at the working parts. They invited their Tibetan friends back to Wisconsin along with several European monks—this time to generate compassion inside an fMRI machine. "The main research question was to see whether some positive qualities such as loving-kindness and compassion or, in general, pro-social altruistic behavior, can be understood as skills and can be trained,"[23] Lutz explained.

Yongey Mingyur Rinpoche was one of the experts who traveled to Wisconsin to think good thoughts inside the scanner. A renowned Buddhist teacher and author of the popular book *The Joy of Living*,[24] he admits that as a young child, he was hardly happy. From the time he was seven or eight until he was thirteen, Mingyur Rinpoche experienced severe panic attacks, despite growing up in a loving family and nurturing spiritual community. He believes that if he had lived in the West, he probably would have been diagnosed with a childhood panic disorder.

The emotional shift for Rinpoche came as a young teenager during a three-year meditation retreat. At first his panic got worse, and he found himself caught in one debilitating anxiety attack after another. He describes that first year on retreat as one of the worst of his life. He became so deeply unhappy, he says, that he "really decided to apply meditation training."[25] At one point he confronted his panic alone in his room for several days, using it as a support for his meditation. After an intensive period of applying what he had learned from his teachers—to relate to his powerful emotions not as his "enemy" or his "boss," as he puts it, but as his friend—the panic was gone. Since that time, he has not experienced another attack.

Rinpoche and the 15 other experienced practitioners taking part in the fMRI study were asked to engage in compassion practice while inside the scanner, as were 16 age-matched novices who had been taught the basics of compassion practice a week before. While inside the scanner, both the monks and the novices heard sounds designed to provoke empathy—like the sounds of a woman in distress—and they heard these sounds while practicing compassion and while in a state of rest. Both groups showed more activation in brain circuits associated with empathy when they were meditating than when they were not, and the monks showed much higher activation in these circuits than the novices. Specifically, long-term practitioners showed heightened activity in the insula—a part of the limbic system that serves as a key interface between physical sensations and emotions—and in the temporal parietal juncture, which appears to play an important role in processing and understanding the emotional states of others. This second area was more active in monks than in novices even during periods of rest, when they were not actively engaged in meditation.[26]

"Both of these areas have been linked to emotion-sharing and empathy," says Davidson. "The combination of these two effects, which was much more noticeable in the expert meditators as opposed to the novices, was very powerful."[27] Lutz and Davidson believe that these observable effects indicate that qualities of mind such as compassion and empathy are trainable, and that this trainability has transformative implications for how we should prepare children for the long emotional road ahead.

"It kind of primes the mind to react with benevolence to others when a situation requires that,"[28] Lutz says. What he saw taking place inside monks' brains matched their temperament and social persona. "Spending time with them is very inspiring," he says, "because they are always very happy—very funny, also."[29]

Teaching happiness

Encouraged by their findings with experienced practitioners, the researchers at UW-Madison wondered if adolescents could be trained to be more spontaneously compassionate. "I think most people would agree," says Davidson, "that this world can use a bit more compassion."[30] He and his team designed an experiment to test if the world could be changed for the better, one teenager at a time.

Teens who signed up for the study took part in compassion training for 30 minutes a day, every day for two weeks. The training was delivered via the Internet—as was a different psychological intervention for a control group (cognitive reappraisal, a form of cognitive therapy). "The cool thing," says Davidson, "was since we delivered the intervention on the Internet, we could monitor exactly how much time they were spending because they had to log on and participate in order to actually do the intervention. We monitored them very closely."[31]

The group receiving compassion training was asked to imagine the suffering of themselves and others—and then to wish for the cessation of that suffering. Traditionally, compassion practice involves the contemplation of phrases like, "May I be free of suffering and the causes of suffering," and "May my mother be free of suffering and the causes of suffering," and so on, focusing attention on loved ones, friends, neutral and then difficult people, and eventually extending the practice to all beings. Teens were instructed to notice visceral sensations—particularly in the area of the heart—as they repeated the phrases and to feel the emotions evoked by the phrases rather than just perform rote repetitions.

After two weeks of practice, teens came into the lab and their brains were scanned while they were shown images of human suffering. The images were not unlike those we commonly see in the news—images of war, for example, or of children suffering from physical deformities. The results of the study are still being analyzed, Davidson says, and their impending publication is likely to set off a cascade of new research into the beneficial neural effects of prosocial emotions and how best to teach them to young people.

What counts as "smart"?

New research shows that social-emotional learning—which includes life skills like regulating negative emotions, cultivating positive ones, and building and maintaining healthy personal relationships—correlates more closely with happiness and life success than any standard measure of academic performance.

"Your grades in school," says Davidson, "your scores on the SAT, mean less for life success than your capacity to cooperate, your ability to regulate your emotions, your capacity to delay your gratification, and your capacity to focus your attention. Those skills are far more important—all the data indicate—for life success than your IQ or your grades."[32]

In 1995, Daniel Goleman's breakthrough book *Emotional Intelligence* laid the evidentiary foundation for social-emotional learning (SEL) curricula, and many educators were convinced. SEL is now a legally mandated component of the K-12 curriculum in several states. "As a professor, I don't like this," Martin Seligman jokes, "but the cerebral virtues—curiosity, love of learning—are less strongly tied to happiness than interpersonal virtues like kindness, gratitude, and capacity for love."[33] The early data showed that kids with strong social-emotional skills were happier and more successful than kids without them—and that they grew up to be happier adults—but it did not prove that the skills were trainable. Strong interpersonal abilities and warm temperament could, conceivably, be written in the genes, so the question remained: Can we teach our kids to better regulate their negative emotions and to cultivate a more positive approach to life?

"The answer is yes," says Davidson. "A very emphatic yes."[34] He points to a recent meta-analysis of 207 studies of SEL programs involving 280,000 kids ages 5–18 from all across the country—from rural, suburban, and urban school settings—which found that social-emotional programs in the schools changed attitudes and behavior for the better *and* improved academic performance. Not only did kids in SEL programs show marked improvements in their social and emotional skills, but they also displayed more positive attitudes, better social and classroom behavior, and less emotional distress and aggression. These gains in social and emotional skills and in indicators of well-being were accompanied by improvements in test scores—both classroom grades and standardized tests. On average, students' achievement scores came up 11 percentile points.[35]

Students in SEL programs, says Goleman, improved "on every measure of positive behavior, like classroom discipline, liking school, and attendance—and went down on rates for every anti-social index, from bullying and fights to suspensions and substance abuse. What's more, there was a drop in numbers of students who were depressed, anxious, and alienated."[36]

Davidson believes this makes sense in light of what we now know about the enormous capacity of children's brains to change in response to experience. The sensitive period for the development of SEL skills is more prolonged than the sensitive period for the development of cognitive skills, because the circuits that underlie SEL skills show enormous plasticity at least through adolescence. Moreover, SEL reduces activity in the brain's distress centers—activity that can impair functioning in prefrontal areas involved in memory, learning, and attention.

When these critical areas of the prefrontal cortex are developing well, adolescents show lower levels of stress hormones that can harm their health in other ways. The anterior cingulate, for example, is associated with the capacity to regulate emotion, and children with more activation in this area have lower levels of cortisol (the primary stress hormone in the body) chugging through their veins, especially late in the evening. Elevated cortisol at bedtime has been shown to wreak havoc on the body by interfering with sleep and disrupting our ability to dream, and it is a risk factor for depression, impaired immune function, and a host of other physical health problems.

Thanks to pictures of minds-in-training, there is little doubt that happiness circuits are highly suggestible, and that they are transformable through intention and effort. "They are among the most plastic circuits in the brain,"[37] Davidson says, and we are becoming increasingly aware that the cultivation of positive, prosocial emotions is vital to that training process.

As we start to draw a map to happiness, it looks as if compassion— for ourselves and for others—is shaping up to be a major landmark.

4

Cooling the flame: pictures of addiction, chronic pain, and recovery

What do a delicious glass of red wine and a nagging backache have in common? Not much, our intuition tells us. One experience (if you enjoy a good red) is something to be savored and something to look forward to at the end of a long, grueling day. The other is something we'll do just about anything to get away from—including, perhaps, drinking too much red wine.

As brain researchers lift the lid on the raw neural experiences of physical addiction and intractable pain, they are discovering that the wars they wage on our bodies and minds can be won, or lost, on surprisingly similar ground. If the glass of red wine becomes addictive, or if the pain becomes chronic, chances are that key areas of the brain have been hijacked by our chemical response to the sensation (whether pleasant or unpleasant), and that these areas—in charge of generating the intense cravings of addiction or the relentless hurt of chronic pain—become overworked and overtrained at the expense of other critical brain functions.

Under normal conditions, our neurochemical responses to sensations are vital to our health and well-being. Evolution picks the quickest studies. We are born knowing to seek experiences that stimulate our neural reward circuitry and to steer clear of hurtful stimuli. It makes a lot of evolutionary sense for us to crave pleasure and avoid pain so that we will learn to eat hearty meals and not run into traffic, but addiction takes that healthy reward-seeking, pain-avoiding behavior to a new level and turns it against us. This is why addiction experts say that drug abuse is something we learn to do very, very well.[1]

Some of us are quicker studies than others, thanks to a combination of genetics and early experiences. An estimated 40 to 60 percent of a person's vulnerability to addiction can be attributed to genetic factors, including the way that our environment influences the expression of our unique genetic identity.[2] Some brains seem more susceptible to the neural imbalances that can lead to addiction and chronic pain, and over time, these conditions can lower our brain metabolism and damage our neurons, shrinking our gray matter at an alarming rate.

That's the dark side of the story. For people suffering from these relentless conditions, the night can feel endless. The good news is that the world is turning. Dr. Nora Volkow, director of the National Institute on Drug Abuse (NIDA), says that thanks to advances in neuroimaging, we now know that drug addiction is a "brain disease that can be treated,"[3] through some combination of medication and behavioral therapy. The same looks true for many types of chronic pain. "There may be a certain control over pain that we don't really realize we have,"[4] says Catherine Bushnell, investigator at McGill University's Center for Research on Pain in Montreal, and evidence out of the new field of "neuroimaging therapy" shows that many chronic pain sufferers can be taught to feel less pain, even to control it altogether. The proof is in the pictures.

Weak character or treatable brain condition?

In 1956, the American Medical Association officially recognized addiction as a disease, but this long-awaited nod by the medical powers-that-be has not stopped the branding of addicts as second-class patients. The mantle of moral failing has been tough to shed, as harsh labels such as "substance abuser" and more pejorative terms like "meth head" and "crack whore" serve so crudely to remind us. Some medical centers still deny alcoholics equal consideration for liver transplants, largely out of fear that the organ will be wasted by alcoholic relapse—a fear invalidated by statistical research. (The Department of Health and Human Services has found that patients who receive a transplant for hepatitis B or C are more likely to suffer disease recurrence and lose the transplanted organ.[5]) Relapse rates among drug addicts are strikingly similar to those of many other chronic diseases, and in a study comparing relapse and noncompli-ance among patients suffering from drug addiction, type I diabetes, hypertension, and asthma, psychologist Thomas McLellan of the Uni-versity of Pennsylvania found that rates of compliance with medica-tion schedules were similar across all four groups.[6]

If this is so, why are addictions often viewed as more hopeless than other illnesses? Addictions can be tough to overcome, but is this because they are somehow hardwired into our basic biology? Do they damage the brain in irreparable ways, or is it possible to change the parts of the brain that are vulnerable to addiction? These questions are as urgent as ever, as each year in this country an estimated 100,000 people succumb to illness or injury due to the use of illicit drugs and alcohol, and more than 440,000 of us die as a result of tobacco use. According to NIDA, substance abuse and addiction carry a crushing national price tag of half a trillion dollars per year.[7]

Addiction experts liken chemical dependency to any disease that disrupts the healthy functioning of the underlying organ—in this case, the brain—and a decade's worth of PET and fMRI scans have provided ironclad proof. Just as the pancreas has stopped working properly in a person with diabetes (another condition in which lifestyle can influence the course of disease), the addicted brain has gone haywire. "We are making unprecedented advances in under-standing the biology of addiction," says David Rosenblum of Boston University School of Public Health. "And that is finally starting to push the thinking from 'moral failing' to 'legitimate illness'."[8]

The news is also promising for chronic pain sufferers—an esti-mated 50 million people in the United States alone, half of whom have been unable to find relief.[9] Many pain sufferers go undiagnosed or untreated; others are told that the problem is "in their heads" because no physical injury can be identified. New findings in pain imaging research, however, mean that pain in the absence of obvious physical injury can no longer be equated with hypochondria or some other quirk of character that might cause people to imagine their dis-comfort. "I think when people say pain is 'all in my head,' it suggests it's not real," Bushnell says. New imaging studies of chronic pain "don't say it's not real, they show that brain activity can create a situa-tion that produces real pain."[10]

Perhaps the most welcome news out of the scanners is that with time and medical care—personalized blends of medication, talk therapy, and potentially neuroimaging therapy, too—many patients will be able to restore balance to the brain's afflicted areas, alleviating physical suffering and starting life over with new freedom.

Your brain on drugs

An egg, cracked into a pan, sizzling ominously—perhaps the most durable imagery in the war on addiction. "This is your brain on drugs," proclaimed the 1987 public service announcement, and the image of that unfortunate, crispy-fried egg was henceforth branded on our national consciousness. It was a memorable ad campaign, to be sure, and a rather crude take on the neural effects of addiction, with a dismal take-home message for anyone already dependent on drugs or alcohol. After an egg is fried, how exactly do you "unfry" it?

Recent imaging research has painted a more sophisticated picture of what happens inside the addicted brain, and just as critically, what happens inside the brain of a person recovering from addiction. In 1996, a pioneering series of PET studies with people treated for cocaine dependence produced two key findings that shaped the course of future imaging research: first, craving cocaine lights up specific neural pathways in the brain (the mesolimbic dopamine system), and second, a cocaine-dependent person's neural response to temptation returns to almost normal within a year. "The highest risk of relapse for cocaine addicts is during the third and fourth week after they've stopped taking the drug," said Dr. Joseph Wu, psychiatrist at UC Irvine and architect of one of the PET studies. "If you can stay abstinent for about a year, you've weathered the periods of greatest vulnerability."[11]

Within five years of these defining discoveries, researchers were using scanning technology to examine addiction with unprecedented precision. Nora Volkow, now director of NIDA, was at Brookhaven National Lab studying the effects of methamphetamine—crystal meth—on the brain. Methamphetamine is a devastating drug that gives the user more than 12 times the rush of dopamine experienced from a natural pleasurable event such as eating good food or having sex. The brain, as it turns out, cannot handle so much bliss. Over time, the drug destroys the pleasure receptors in the brain and creates serious deficits in judgment, making relapse much more likely—one of the reasons why meth use has reached epidemic proportions in this country. By 2007, more than 13 million people aged 12 or older—that's more than 5 percent of the teen and adult population—had tried, or had become addicted to, the corrosive drug.[12]

Dr. Volkow and colleagues began documenting the neural effects of crystal meth in 2001, and they concluded that our pleasure response to the chemical cuts a wide swath of destruction through the brain, damaging dopamine receptors, lowering levels of dopamine transporters, and hampering glucose metabolism in the orbitofrontal cortex (a part of the brain linked to impulse control).[13] Low metabolism in this brain region had already been observed in people addicted to cocaine and alcohol; this was the first study to show that the problem was echoed, and amplified, with meth use. Meanwhile, an fMRI study with meth addicts at UC San Diego linked lower activity in the orbitofrontal and prefrontal cortices with significant impairments in cognition and judgment.[14]

For people whose lives have been eviscerated by crystal meth addiction and for their loved ones praying for recovery, however, there was also cause to celebrate. Volkow and colleagues demonstrated the brain's extraordinarily resilient ability to heal after long-term abstinence, even from a drug as toxic as meth. After 14 clean months, PET scans showed that most of the damaged dopamine receptors had come back online, though significant deficits in judgment, memory, and motor coordination were still present.[15] Dr. Richard Rawson, associate director of UCLA's Integrated Substance Abuse Programs, said of this new evidence, "It's a wonder any meth users ever get better, but in fact they do."[16]

Researchers emphasize that some deficits might be permanent, especially after heavy, prolonged meth use. The first high-resolution MRI study of people with decade-long addictions to meth showed a "forest fire of brain damage," according to lead UCLA investigator Paul Thompson.[17] The limbic region had lost 11 percent of its tissue, and the hippocampus—critical to the formation of new memories—had lost 8 percent. These deficits are similar to neuronal losses seen in early Alzheimer's, and suggest that early intervention might be much more critical to recovery from meth addiction than from many other common chemical dependencies.

Meth is a devastating drug, but hardly the chief addictive killer in the United States. Cigarettes account for nearly one in five deaths, killing more Americans than alcohol, illegal drugs, car accidents, suicide, AIDS, and homicide combined. The CDC reports that 43.4 million U.S. adults were smokers in 2007—nearly 20 percent of all adults.[18]

Imaging research recently linked smoking addiction to the insula, a small brain region that also figures highly in fMRI studies of compassion and empathy. This region is now believed to be a vital interface between physical sensations and emotions, and a January 2008 study made it the new darling of addiction research. MRI scans of stroke victims showed that those patients who found it strangely easy to quit smoking had suffered damage to the insula—exposing its key role in the circuitry of addiction.

"There's a whole neural circuit critical to maintaining addiction, but if you knock out this one area, it appears to wipe out the behavior,"[19] said Dr. Antoine Bechara, neuroscientist at the Brain and Creativity Institute at the University of Southern California and a senior author on the paper. Out of 32 former smokers who had suffered brain injuries, 16 quit easily and no longer had cravings. The MRI scans showed that these 16 patients were far more likely to have suffered injury to the insula than to any other part of the brain.

"This is the first time we've shown anything like this, that damage to a specific brain area could remove the problem of addiction entirely," said Volkow. "It's absolutely mind-boggling."[20] Her agency, NIDA, financed the study, along with the National Institute of Neurological Disorders and Stroke. A Cinderella story of neuroscience, the insula was overlooked for decades partly because it is buried so deep in our heads that it is nearly impossible to study using surface electrodes. Long believed to be a primitive part of the brain, the insula was implicated in basic biological functions such as eating and having sex, but this unassuming bundle of neurons has turned out to be a major operations hub for much of the neural activity we associate with human nature. The frontal insula, in particular, is more highly developed in humans and apes than in other animals, and it is key to sensing our own emotions and the emotions of others. It is now thought to be the stage upon which bodily sensations are cast as emotions and revealed to the rest of the brain. Dr. Antonio Damasio, director of the Brain and Creativity Institute and one of the early champions of the insula's central importance to human cognition, was not surprised to learn that it was a key piece of the addiction puzzle. "It is on this platform," he said, "that we first anticipate pain and pleasure, not just smoking but eating chocolate, drinking a glass of wine, all of it."[21]

Brain imaging might be our x-ray vision into the nature of addiction, but what will we do with what we see? How will our newfound knowledge affect the way we treat people with substance abuse disorders? The implications are radical because images of addiction recast the problem as hopeful rather than hopeless, treatable rather than terminal, and because they shatter stigmas attached to these and other brain disorders—stigmas that too often cause sufferers to hide their conditions instead of seeking treatment they so desperately need.

"The future is clear," says Volkow. "In ten years we will be treating addiction as a disease, and that means with medicine."[22] Some drugs have already shown great promise in preventing relapse—acamprosate for alcohol addiction, naltrexone for alcohol and heroin addiction, for example—and more than 200 other compounds are in the research pipeline. Some of the larger pharmaceutical companies have shown reluctance to develop drugs to treat addiction, choosing instead to associate themselves with more socially acceptable diseases like high cholesterol or arthritis. "I have been imploring the bigger companies to work on this," says Volkow. "Their scientists get it, but the business people are tough to persuade."[23]

That tide is turning, says Steven Paul, the head of research for Eli Lilly, who notes that there used to be a stigma attached to drugs to treat depression, too, and then along came Prozac. "Anything that has a large unmet need," says Paul, "is ultimately going to succeed commercially."[24]

Another treatment showing promise is not a drug at all, but rather a state of mind. Scientists are looking at whether people might be able to overcome their addictions by consciously modulating activity in the insula and other targets in the brain's reward circuitry. Momentum for this work has come from an unexpected place: imaging research with chronic pain patients.

Your brain on pain

Even if we don't suffer the debilitating effects of chronic pain ourselves, we probably love someone who does. Perhaps it's a father who can't sleep because of the crippling neuropathy in his feet, or a friend who can't lift her toddler because her back never healed after pregnancy. For millions of chronic pain sufferers in this country alone, a fix can be elusive: The American Pain Foundation has found that only one in four pain sufferers receives proper treatment, while the National Sleep Foundation estimates that one in three American adults loses more than 20 hours of sleep each month due to pain.[25]

In 2004, brain researchers made headlines by revealing that long-term pain actually shrinks the brain of the person who is hurting. MRI scans showed that the gray matter volume of people who suffer from back pain for more than a year is 5 to 11 percent less than that of healthy volunteers, and—perhaps most tellingly—that the length of time a person has suffered correlates directly with the amount of tissue loss.[26] "The magnitude of this decrease is equivalent to the gray matter volume lost in 10 to 20 years of normal aging,"[27] the researchers noted, and their findings underscore the importance of looking at pain—even extremely localized pain—as a sign of an injured physical system, and the brain as a critical and vulnerable part of that system.

These were stunning results, with dire implications for chronic pain patients, but researchers were no closer to understanding the injury mechanism. How might relentless physical pain actually be harming brain tissue? In February 2008, there was another breakthrough, one that pointed to reasons why chronic pain sufferers often struggle with other debilitating problems such as depression, anxiety, and trouble focusing and making decisions. Researchers at Northwestern University's Feinberg School of Medicine wondered how the brain of a pain sufferer might function differently than a pain-free person's, and to find out, they scanned the brains of chronic low-back-pain sufferers and healthy people while they performed a simple visual task—tracking a moving bar across a computer screen. Both groups performed the work well, but fMRI images showed that pain sufferers achieved that state of concentration by using their brains quite differently than the pain-free group.

Activity in healthy brains, the investigators noted, tends to shift smoothly back and forth between task-related areas and the so-called default or resting-state network, which encompasses areas in the frontal, parietal, and medial temporal lobes. This network is thought to play important roles in learning and memory and in maintaining energy equilibrium in the brain. Also referred to as the "dark network," these areas buzz with activity when we appear to be doing nothing at all, and go as cold and black as deep space when we are engaged in a task or solving a problem—when we're mentally connected to the moment, rather than time-traveling through our past or projecting ourselves onto an imagined future.

In the pain-free subjects, these areas go as dark as predicted whenever other areas of the brain come online to tackle the task at hand. In chronic pain sufferers, however, a front region of the cortex mostly associated with emotion "never shuts up," said Dr. Dante Chialvo, lead investigator on the study and associate professor of physiology at the Feinberg School. "If you are a chronic pain patient, you have pain 24 hours a day, seven days a week, every minute of your life," Chialvo said. "That permanent perception of pain in your brain makes these areas in your brain continuously active."[28] Such relentless activity, in turn, can create wear and tear on neurons and their connections, perhaps even causing permanent damage. "We know when neurons fire too much they may change their connections with other neurons or even die because they can't sustain high activity for so long,"[29] Chialvo explained.

For those of us of a more stoic bent, the take-home message is clear: Quit grinning and bearing it. For our more emotionally effusive friends, catastrophizing won't solve the problem either. We need to fix the broken system, but how? For many chronic pain sufferers, some combination of physical therapy and medication helps, and for others, alternative therapies such as massage, acupuncture, and mindfulness meditation might greatly alleviate symptoms. But what about the millions of people who feel that they have exhausted their options without touching their pain?

An entirely new approach, dubbed by its inventors "neuroimaging therapy," might provide welcome answers.

Mind control

In some contexts, the idea that our mental activity affects our sensory experiences seems unsurprising. We think sexy thoughts and increase our arousal; we anticipate a dreaded exam or public speaking event and our stomach bungee-jumps. Something about the experience of pain, however, feels—well, much too *real* to be affected by anything happening in our minds. When we feel pain, we believe it, just as we believe the floor under our feet or an apple on our plate. Real physical phenomena have real sensory attributes, such as hard and smooth, crisp and sweet, stabbing and burning. *Not* to believe in pain as a solid, observable fact, just like the floor or the apple, feels like denial. But what if it turns out to be entirely possible to change our experience of pain—even feel less of it—just by adjusting our mental activity?

This is the idea behind a revolutionary clinical trial underway in Menlo Park, California. Sean Mackey, director of the Neuroimaging and Pain Lab at Stanford University, and neurophysiologist Christopher deCharms, chief executive of the California company Omneuron, designed a study to see if chronic pain patients could be taught to use visual fMRI feedback to regulate their experience of pain. People have successfully used much simpler biofeedback technology (EKG, for example) to control physiological markers such as heart rate, but until now, pain has been trickier. It involves a variety of brain and body systems and has never been distilled down to one simple physiological indicator.

Now fMRI technology gives us a direct window into the parts of the brain in charge of saying "ouch." For the purposes of this study, investigators focused on an area of the brain called the rostral anterior cingulate cortex (rACC), which is believed to be key to pain perception and regulation and to other critical cognitive processes such as attention, emotion, and executive function. "There is an interesting irony to pain," deCharms says. "Everyone is born with a system designed to turn off pain. There isn't an obvious mechanism to turn off other diseases like Parkinson's. With pain, the system is there, but we don't have control over the dial."[30] The purpose of the study was, in effect, to hand chronic pain patients the remote control.

Twelve patients suffering from intractable pain, eight men and four women, were selected from the Stanford University Pain Management Service. (Eight took part in a study group; four in a patient control group.) Their pain had already lasted an average of three-and-a-half years and had proven resistant to other treatments. Once they were inside the scanner, study participants were instructed to employ several strategies to control their pain, such as focusing their attention on the pain rather than avoiding it; attempting to perceive a painful stimulus as neutral rather than "tissue-damaging, frightening, or overwhelming;" and trying to perceive the stimulus as mild rather than high intensity.

The eight patients in the experimental group were shown a computer-generated image of a flame, the intensity of which represented the strength of activation of the rACC, and they were instructed to increase or decrease its intensity with the strategies they had learned. The results were persuasive: After training, patients in the real-time fMRI study group felt their pain reduced by 64 percent on average, with five of eight patients reporting that their pain was reduced by at least half. Moreover, ramped-down rACC activation correlated significantly with pain reduction reported by patients.[31]

"It is the mind-body problem—right there on the screen," says deCharms. "We are doing something that people have wanted to do for thousands of years. Descartes said, 'I think, therefore I am.' Now we're watching that process as it unfolds."[32]

To distinguish between fMRI-induced learning and other learning or placebo effects, several separate healthy control groups (in addition to the patient control group) were trained and tested using similar procedures, but in the absence of valid fMRI feedback. One of the healthy groups received identical instructions on how to regulate their pain (induced by a hot metal probe on the skin) with no fMRI feedback; another group received identical instructions but was shown real-time fMRI information from an area of the brain unrelated to pain processing. Still another group was shown the fMRI input from the person whose brain had been scanned just prior to their own, whereas the patient control group received autonomic biofeedback information (such as heart rate and blood pressure) instead of fMRI input. No control group saw the reduction in pain achieved by the study group, and pain reduction for the study group was three times greater than for the patient group receiving autonomic biofeedback.

"It is particularly interesting," the investigators noted, "that pain patients need to be able to observe the functioning of the brain's pain system to learn this form of control over these systems because pain patients already have continuously available sensory feedback of their own pain level, they already have a strong motivation to learn to control their pain, and they typically have tried and practiced many strategies to alleviate their pain over many years."[33]

Is there something special about the brains of some chronic pain patients that makes pain more resistant to less targeted types of therapy? Similar injuries result in wildly different outcomes for different patients, and Mackey, deCharms, and colleagues speculate that this might be because some patients' neural pain-control systems are more engaged or more effective than others'. Neuroimaging therapy, they believe, might provide a different route for those of us not so naturally blessed, by teaching "up-regulation of the pain control system based on targeted neuroplasticity through training"[34]—a crash course in pain control. "I believe the technique may make lasting changes because the brain is a machine designed to learn,"[35] says deCharms.

What powers might this approach hold for other chronic problems involving well-delineated brain systems, like addiction? Omneuron is currently looking at whether smokers can learn to train their brains to down-regulate their craving circuitry, just as pain sufferers can down-regulate their perception of pain. "We hope to develop a novel therapeutic for addiction, which will create a new way of treating these patients," says deCharms. "We're just in the early stages of this research, but the hope is that rather than using purely pharmacologic or cognitive [therapy] approaches, this would provide another avenue to treat this major disorder."[36]

Nora Volkow says she has been intrigued by this potential application of real-time fMRI since deCharms and colleagues published their report on chronic pain. (That research was funded, in part, by her agency NIDA.) "Wouldn't it be wonderful if we could train the brain to regulate the response to craving?"[37] she says. There are plenty of potential snags for this approach, such as whether cooling the activity of the insula to calm cravings might produce other, unwanted behavioral changes. The insula might be where the insatiable cravings of addiction have taken up residence, but it is also the seat of positive social emotions like love, empathy, and compassion—a collection of humanity's greatest hits.

Even if neuroimaging therapy works for addiction without undesirable side effects, there will be other serious challenges to employing it clinically. For starters, how will we carry its benefits with us outside the scanner? "One caveat is whether the exercises done with real-time fMRI can then be extended to real-life situations," says Rita Goldstein, brain researcher at Brookhaven National Laboratory. "But the prediction is that it will be like exercise, strengthening the functions subserved by those specific brain regions."[38]

If this novel application of fMRI technology produces lasting effects for people suffering from addiction and chronic pain, it will mean that watching our brains self-regulate can provide a lifesaving shortcut to neuroplastic changes—an express ticket on the road to recovery.

The pictures

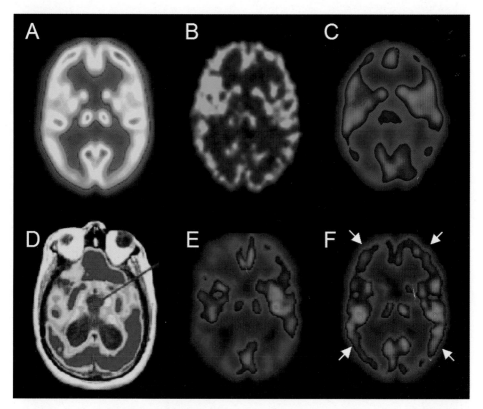

PET images of healthy brain metabolism and of brain metabolism in the vegetative state. *These images illustrate that glucose metabolism in the conscious waking state (A) is about twice as high as in states of altered wakefulness due to general anesthesia (B) and deep sleep (C). A patient in a vegetative state due to traumatic injury (D) shows close-to-normal overall brain metabolism, whereas a vegetative patient whose injury is due to carbon monoxide poisoning (E) recovers full awareness (F) with little increase in global metabolic function.* Courtesy of Steven Laureys, from S. Laureys, "The Neural Correlates of (Un)awareness: Lessons from the Vegetative State," *Trends in Cognitive Sciences* 9, no. 12 (December 2005): 556–559, http://www.sciencedirect.com/science/journal/13646613. Copyright 2005, reprinted with permission from Elsevier.

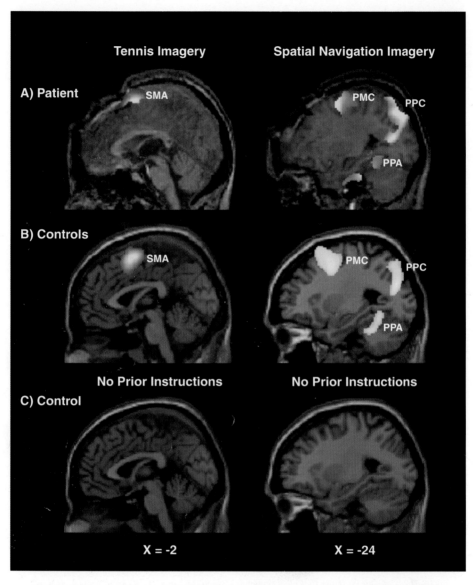

A patient diagnosed as vegetative shows remarkable brain activity on fMRI scans. Her neural activity is indistinguishable from healthy volunteers during mental imagery tasks. These images show (A) the patient's brain activity in regions involved in imagining coordinated movements and spatial navigation while she imagines playing tennis and moving around a house, (B) statistically indistinguishable activity in these brain regions in healthy volunteers asked to perform the same imagery tasks, and (C) the result when one healthy volunteer undergoes the same fMRI procedure as in (A) and (B), but this time hearing noninstructive sentences such as, "The man played tennis," or "The man walked around his house." Courtesy of Adrian Owen, from A.M. Owen et al., "Response to Comments on 'Detecting Awareness in the Vegetative State,'" *Science* 315 (2 March 2007): 1221, http://www.sciencemag.org/cgi/content/full/315/5816/1221c. Reprinted with permission from the American Association for the Advancement of Science (AAAS).

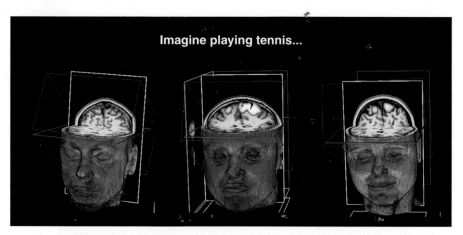

Imagine playing tennis...

Three healthy volunteers imagine playing tennis. *All show similar activity in the supplementary motor area. This composite image was created by superimposing functional MRI data on 3-D anatomical structural MRI.* Courtesy of Adrian Owen, from A.M. Owen et al., "Using Functional Magnetic Resonance Imaging to Detect Covert Awareness in the Vegetative State," *Archives of Neurology* 64, no. 8 (August 2007): 1098–1102. Copyright 2007 American Medical Association. All rights reserved. Reprinted with permission.

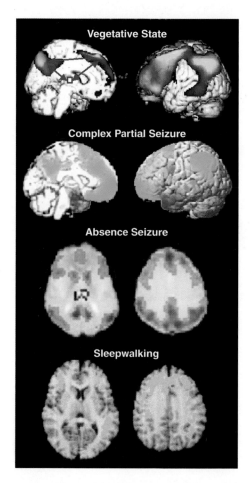

Vegetative State

Complex Partial Seizure

Absence Seizure

Sleepwalking

Images of the vegetative state and other states devoid of awareness. *A common characteristic of the vegetative state is decreased blood flow to a widespread neural network encompassing prefrontal and parietal regions. Other studies of similar, but transient, states where wakefulness is present without awareness also show reduced blood flow in this frontoparietal network. In these images of complex partial seizures, reduced blood flow appears in green; in absence seizures, reduced blood flow appears in blue; and in sleepwalking, reduced blood flow appears in yellow.* Courtesy of Steven Laureys, from S. Laureys, "The Neural Correlates of (Un)awareness: Lessons from the Vegetative State," *Trends in Cognitive Sciences* 9, no. 12 (December 2005): 556–559, http://www.sciencedirect.com/science/journal/13646613. Copyright 2005, reprinted with permission from Elsevier.

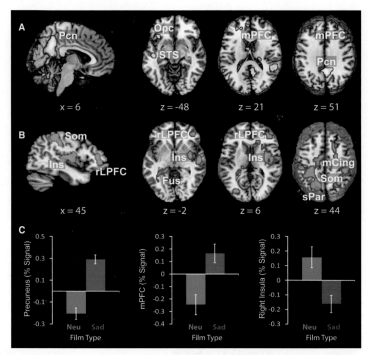

The neural expression of sadness. *Areas of heightened (A) and decreased (B) brain activity on fMRI scans in response to sad film clips in people who had signed up for—but who had not yet attended—a mindfulness-based stress reduction course. The graphs in (C) compare activity in three brain regions of interest while subjects watched neutral and sad film clips.* Courtesy of Norman Farb and Adam Anderson, from N.A.S. Farb et al., "Minding One's Emotions: Mindfulness Training Alters the Neural Expression of Sadness," *Emotion* (in press). Copyright 2009 American Psychological Association. Reprinted with permission.

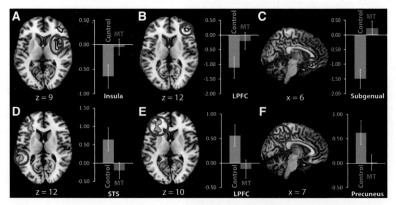

Different mental approaches to dealing with sadness. *Differences in regional activation and deactivation between people who were scheduled to attend, and people who had already attended, a mindfulness-based stress reduction course (the "control" group and the "mindfulness training" or "MT" group, respectively) in response to sad film clips. The top panel shows sadness-related deactivations on fMRI scans in the control group as compared to non-significant activity in the mindfulness training group, while the bottom panel shows sadness-related activations in the control group as compared (again) to non-significant activity in the MT group.* Courtesy of Norman Farb and Adam Anderson, from N.A.S. Farb et al., "Minding One's Emotions: Mindfulness Training Alters the Neural Expression of Sadness," *Emotion* (in press). Copyright 2009 American Psychological Association. Reprinted with permission.

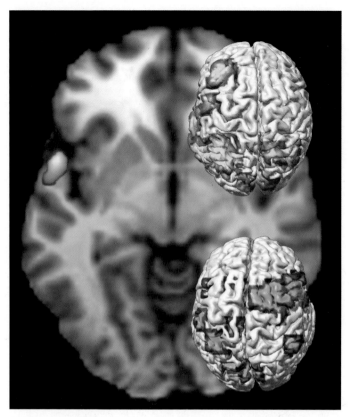

The emotional brain. *This image shows regions of the brain that correlate with positive and negative emotions (as measured by a standard scale) in a large group of volunteers. The orange areas show the correlation between glucose metabolism and positive emotion (PET images superimposed on a high-resolution MRI image), whereas the lower image—the 3D-rendered "blue" brain—shows regions that correlate with negative emotions.* Courtesy of Richard Davidson and Terry Oakes.

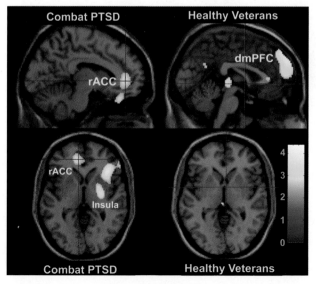

Imaging combat PTSD. *These PET images show neural differences in response to reminders of trauma in healthy veterans and in veterans suffering from combat PTSD. In veterans with PTSD, stress hormone responses correlate with more pronounced activity in neural circuits involved in experiencing pain and negative emotion, whereas in veterans without PTSD, responses occur in circuits that are key to emotion regulation.* Courtesy of Anthony King.

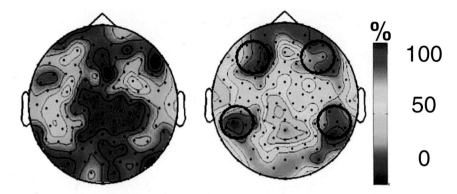

Heightened gamma power after mental training. *These EEG images show the scalp distribution of gamma activity during meditation. The color scale indicates the percentage of subjects in each group that had an increase in gamma activity during mental training. (Controls are on the left; long-term practitioners are on the right). Gamma activity is believed to be an indicator of neural synchrony—the integration of localized neural processes into higher-order cognitive and emotional functions. For the long-term practitioners, relative gamma activity correlated with the total number of hours (in the tens of thousands) that they had practiced.* Courtesy of Richard Davidson, from A. Lutz et al., "Long-Term Meditators Self Induce High-Amplitude Gamma Synchrony During Mental Practice," *PNAS* 101, no. 46 (16 November 2004): 16369–16373. Copyright 2004 National Academy of Sciences, U.S.A. Reprinted with permission.

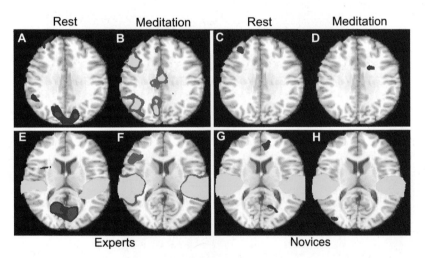

Training the brain to be more compassionate. *These fMRI images show differences between the neural responses of expert meditators and novices to emotional sounds. Inside the scanner, both groups heard sounds designed to provoke empathy—like the sounds of a woman in distress—and they heard these sounds while practicing compassion and while resting. Both groups showed more activation in brain circuits associated with empathy when they were meditating than when they were not, and the monks showed much higher activation in these circuits than the novices. Colored areas show a negative (blue) or positive (orange/yellow) impulse response across ten seconds to all emotional sounds for meditation novices and experts.* Courtesy of Richard Davidson, from A. Lutz et al., "Regulation of the Neural Circuitry of Emotion by Compassion Meditation," *PLoS ONE* 3, no. 3 (26 March 2008): e1897. Reprinted under a Creative Commons Attribution License.

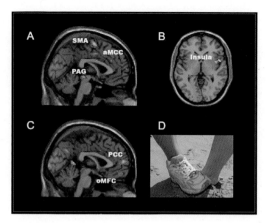

Natural empathy. *When children observe others in pain (whether accidentally or intentionally inflicted), blood flow increases in brain regions involved in the first-hand experience of pain (A and B). When children observe someone intentionally harming another person (as in D), regions involved in representing social interactions and the intentions of others also become active (C).* Courtesy of Jean Decety, from J. Decety et al., "Who Caused the Pain? An fMRI Investigation of Empathy and Intentionality in Children," *Neuropsychologia* 46 (2008): 2607-2614, http://www.sciencedirect.com/science/journal/00283932. Copyright 2008, reprinted with permission from Elsevier.

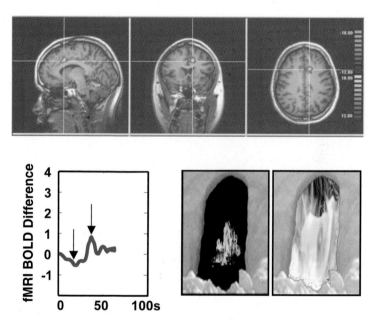

Learned control over pain perception and regulation. *These fMRI images show a group of chronic pain patients learning to control activation in a target brain region (the rACC) thought to be central to pain perception and regulation. The patients learned to control activity in this brain region by viewing a realtime, scrolling line graph of fMRI feedback from the region (lower left) and a video depicting the same information as a smaller or larger virtual flame (lower center and right) depending on the strength of the fMRI signal. (The two sample flame images shown here correspond to the two arrows on the fMRI signal graph.) The colored regions on the brain scans show the change in activation in the rACC, comparing the last training session to the first.* Courtesy of Christopher deCharms, from R.C. deCharms et al., "Control over Brain Activation and Pain Learned by Using Real-Time Functional MRI," *PNAS* 102, no. 51 (20 December 2005): 18626–18631. Copyright 2005 National Academy of Sciences, U.S.A. Reprinted with permission.

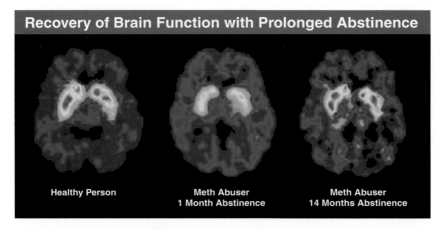

Recovery of brain function after prolonged abstinence from drug abuse. *These PET images show the brain's remarkable ability to heal, at least partially, from drugs—in this case, the corrosive and extraordinarily addictive drug methamphetamine (crystal meth). The red areas show recovery of dopamine transporters after 14 months of abstinence.* Courtesy of Nora Volkow, from N.D. Volkow et al., "Loss of Dopamine Transporters in Methamphetamine Abusers Recovers with Protracted Abstinence," *Journal of Neuroscience* 21, no. 23 (1 December 2001): 9414–9418. Copyright 2001 Society for Neuroscience. Reprinted with permission.

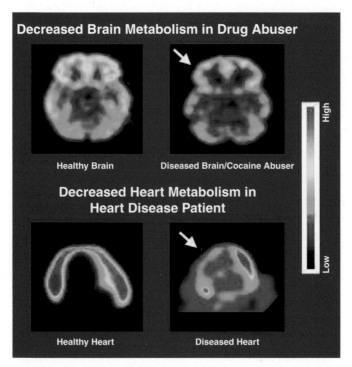

Drug addiction is a brain disease. *These images of brain and heart metabolism in addiction and heart disease show that addiction is like other diseases that disrupt the normal, healthy functioning of the underlying organ.* Courtesy of Nora Volkow, from the National Institute on Drug Abuse, *Drugs, Brains, and Behavior: The Science of Addiction*, NIH Pub. No. 07–5605 (2007). Reprinted with permission.

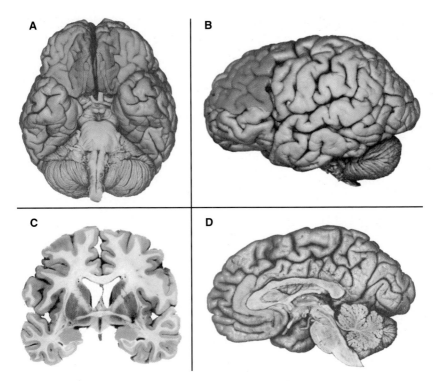

Regulating our most powerful emotions. *These images show several brain areas that are key to emotion regulation: (A) orbital prefrontal cortex (green) and ventromedial prefrontal cortex (red), (B) dorsolateral prefrontal cortex, (C) amygdala, and (D) anterior cingulate cortex. Irregularities in these structures—or in the connections among them—are associated with malfunctions in emotion regulation and with abnormal tendencies toward aggression and violence.* Courtesy of Richard Davidson, from R.J. Davidson et al., "Dysfunction in the Neural Circuitry of Emotion Regulation—A Possible Prelude to Violence," *Science* 289 (28 July 2000): 591–594, http://www.sciencemag.org/cgi/content/abstract/sci;289/5479/591. Reprinted with permission from AAAS.

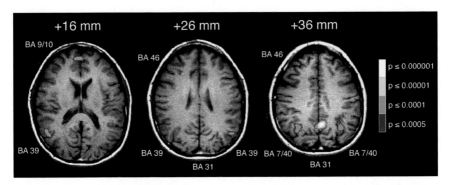

Moral judgments are different when it's personal. *Three axial slices of the brain show disparities in neural activity between people wrestling with moral dilemmas that feel "up close and personal," and people considering nonmoral dilemmas or moral dilemmas that feel impersonal. Areas associated with emotion are more active in dilemmas that feel personal, whereas areas associated with working memory are recruited for moral-impersonal and nonmoral cases.* Courtesy of Joshua Greene, from J.D. Greene et al., "An fMRI Investigation of Emotional Engagement in Moral Judgment," *Science* 293 (14 September 2001): 2105–2108, http://www.sciencemag.org/cgi/content/abstract/293/5537/2105. Reprinted with permission from AAAS.

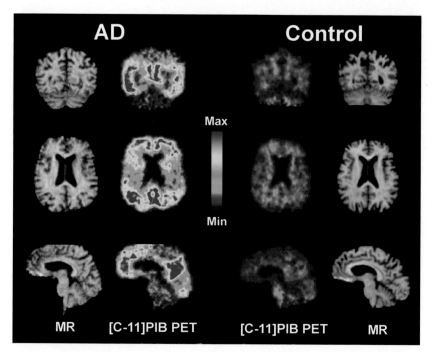

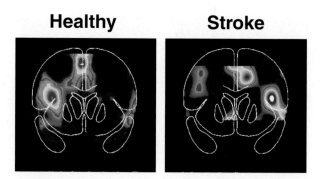

New tools for diagnosing Alzheimer's. *On the far left are MRI scans of a patient with Alzheimer's disease viewed as if looking face-on (top image), from the top of the head down (middle image), and from the side (bottom image). The far right shows similar MRI images from a healthy elderly person with no memory impairment. Next to the MRI scans are the corresponding PET scans obtained by using a marker for amyloid plaques called Pittsburgh Compound-B (PiB). The red, orange, and yellow areas show regions with heavy plaque formation in the patient (red indicates the highest levels). The new findings, which correlate plaque deposits from living patients to their post-mortem autopsy results, will ultimately aid in the early diagnosis, monitoring, and treatment of Alzheimer's. Until now, Alzheimer's could only be definitively diagnosed on autopsy.* Courtesy of William Klunk and the University of Pittsburgh PET Amyloid Imaging Group.

Recovery after stroke. *A 72-year-old man suffered a stroke that damaged a portion of his left frontal lobe. Remarkably, he was able to perform many language and speech tasks, which, in healthy young people (left image), involve the part of the left frontal lobe that had been damaged. His preserved abilities led investigators to guess that he might be using alternative brain areas to compensate for his injury. To test their theory, they imaged his brain using positron emission tomography (PET). They found that speech generation tasks now activated his right frontal cortex (right image) rather than his left.* Courtesy of Randy Buckner, from R.L. Buckner et al., "Preserved Speech Abilities and Compensation Following Prefrontal Damage," *PNAS* 93 (February 1996): 1249–1253. Copyright 1996 National Academy of Sciences, U.S.A. Reprinted with permission.

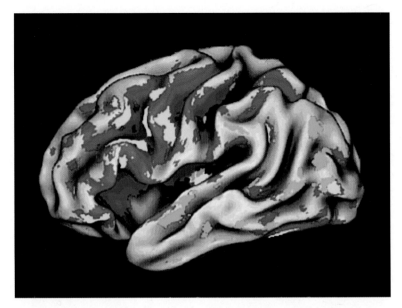

Alzheimer's disease may leave some kinds of memory unharmed. *This fMRI study of people living with Alzheimer's demonstrates that frontal brain regions used during a word classification task, shown in red, reduce their activity with practice, as shown in yellow. Frontal regions of the brain are involved in high-level cognition and planning, and the finding that these regions change activity with learning suggests that certain memory functions affecting complex cognitive abilities are preserved in Alzheimer's.* Courtesy of Randy Buckner.

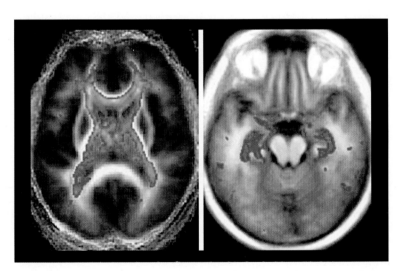

Alzheimer's disease is distinct from normal age-related brain changes. *Different MRI methods reveal changes associated with normal aging, which occur in frontal regions, and those prominent in Alzheimer's disease, which affect the hippocampus. Areas of color show regions of change for normal aging (left image) and for Alzheimer's disease (right image).* Courtesy of Randy Buckner, from R.L. Buckner, "Memory and Executive Function in Aging and AD: Multiple Factors that Cause Decline and Reserve Factors that Compensate," *Neuron* 44 (30 September 2004): 195–208, http://www.sciencedirect.com/science/journal/08966273. Copyright 2004 Cell Press, reprinted with permission from Elsevier.

Why remembering can feel like reliving. *This fMRI study of perception and remembering shows that areas of the sensory cortex are reactivated during recall of vivid sounds (top) and pictures (center). Yellow areas in the bottom images show the overlap in brain activity during perception and remembering of vivid sensory input.* Courtesy of Mark E. Wheeler.

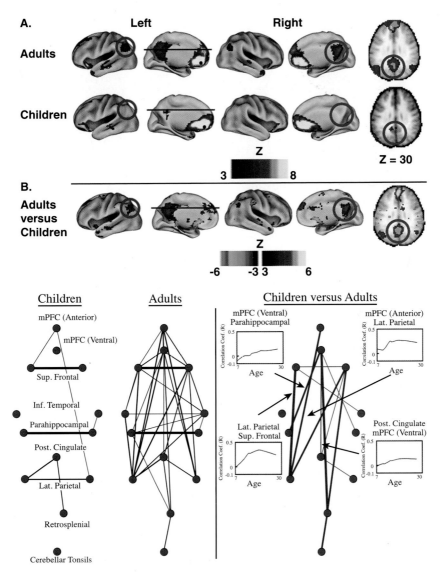

A.

Left   Right

Adults

Children

Z

3   8

Z = 30

B.

Adults
versus
Children

Z

-6   -3 3   6

Children   Adults   Children versus Adults

mPFC (Anterior)

mPFC (Ventral)

Sup. Frontal

Inf. Temporal

Parahippocampal

Post. Cingulate

Lat. Parietal

Retrosplenial

Cerebellar Tonsils

mPFC (Ventral)
Parahippocampal

Correlation Coef. (R)
0.5
0
-0.1 7   30
Age

mPFC (Anterior)
Lat. Parietal

Correlation Coef. (R)
0.5
0
-0.1 7   30
Age

Lat. Parietal
Sup. Frontal

Correlation Coef. (R)
0.5
0
-0.1 7   30
Age

Post. Cingulate
mPFC (Ventral)

Correlation Coef. (R)
0.5
0
-0.1 7   30
Age

The evolving default network—childhood to adulthood. *These images compare three maps of functional connections within the brain: an adult map, a childhood map, and a map of the direct comparison of connections in adults' and children's brains between the ventromedial prefrontal cortex (marked by the solid black dot)—a central constituent of the default, or "dark," network—and other areas of this network. The default network is thought to be central to an autobiographical sense of self and to stimulus-independent, non-goal-directed thought. Functional connections with other areas of the default network (highlighted by blue circles) are present in adults but virtually absent in children. The bottom images are graphic representations of functional connectivity data, showing that default regions are only sparsely connected in children and are highly integrated in adults (line width is proportional to connection strength). In the "Children versus Adults" comparison, blue lines represent significantly greater connectivity in adults than in children.* Courtesy of Damien Fair, from D.A. Fair et al., "The Maturing Architecture of the Brain's Default Network," *PNAS* 105, no. 10 (11 March 2008): 4028–4032. Copyright 2008 National Academy of Sciences, U.S.A. Reprinted with permission.

Novice Group

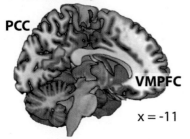

PCC

VMPFC

x = -11

Narrative > Experiential

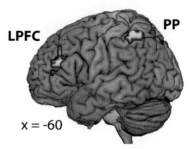

LPFC

PP

x = -60

Experiential > Narrative

MT Group

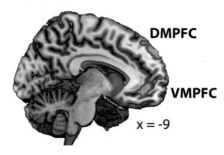

DMPFC

VMPFC

x = -9

Narrative > Experiential

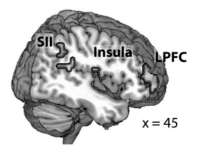

SII

Insula

LPFC

x = 45

Experiential > Narrative

Mindfulness training disentangles awareness of the self in the moment from the extended sense of self across time. After eight weeks of mindfulness training, study participants showed much greater activation in the right insula—an area that has been tightly linked to awareness of body sensations—during experiential self-focus, whereas a strong coupling of the medial prefrontal cortex (mPFC) and right insula that was observed in the untrained group disappeared in the mindfulness group. These findings suggest that training in present-centered awareness might encourage a shift away from processing sensory information through the lens of the narrative self (as supported by activity in the mPFC), and that two distinct forms of self-reference that are habitually integrated can be dissociated through mental training. Brain areas linked to narrative self-focus are shown in blue, and areas linked to experiential self-focus are shown in yellow/red. Courtesy of Norman Farb.

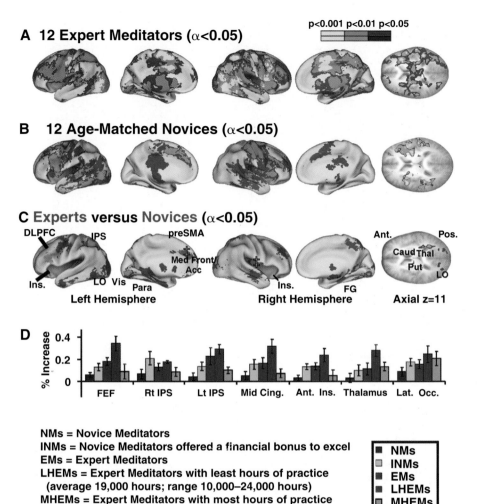

A 12 Expert Meditators (α<0.05)

p<0.001 p<0.01 p<0.05

B 12 Age-Matched Novices (α<0.05)

C Experts versus Novices (α<0.05)

DLPFC   IPS   preSMA   Ant.   Pos.

Med Front/Acc   Caud Thal

Ins.   LO   Vis   Para   Ins.   FG   Put   LO

Left Hemisphere   Right Hemisphere   Axial z=11

D

% Increase

0.4

0.2

0

FEF   Rt IPS   Lt IPS   Mid Cing.   Ant. Ins.   Thalamus   Lat. Occ.

NMs = Novice Meditators
INMs = Novice Meditators offered a financial bonus to excel
EMs = Expert Meditators
LHEMs = Expert Meditators with least hours of practice
  (average 19,000 hours; range 10,000–24,000 hours)
MHEMs = Expert Meditators with most hours of practice
  (average 44,000 hours; range 37,000–52,000 hours)

■ NMs
□ INMs
■ EMs
■ LHEMs
□ MHEMs

Experts in attention and emotion regulation. *These fMRI images show that expert medita-tors with an average of 19,000 hours of practice have more activation in brain regions associ-ated with sustained attention, whereas expert meditators with an average of 44,000 hours have less activation, suggesting that sustained practice ultimately leads to a level of facility requiring less effort. When distracted by sounds, expert meditators showed less activation in regions asso-ciated with discursive thoughts, daydreams, and emotions. These images show (A) brain activa-tion in concentration meditation versus resting state for 12 expert meditators, (B) brain activation in concentration meditation versus resting state for 12 age-matched novice medita-tors, (C) comparison of expert and novice meditators (red hues reflect greater activation in experts versus novices; blue hues reflect greater activation in novices versus experts), and (D) activation in regions of interest for attention.* Courtesy of Richard Davidson, from J.A. Brefczynski-Lewis et al., "Neural Correlates of Attentional Expertise in Long-Term Meditation Practitioners," *PNAS* 104, no. 27 (3 July 2007): 11483–11488. Copyright 2007 National Academy of Sciences, U.S.A. Reprinted with permission.

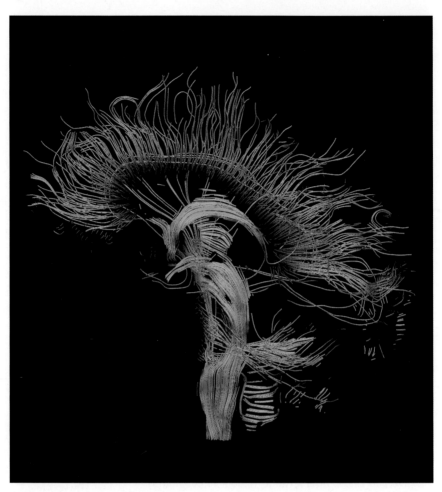

The connected brain. *Diffusion MRI methods enable us to visualize the "wiring," or connections, within the brain, creating images based on the direction of water diffusion along the long, thin nerve fibers (axons) that are the major constituent of white matter.* Image rendered by Thomas Schultz, 22 September 2006; dataset courtesy of Gordon Kindlmann at the Scientific Computing and Imaging Institute, University of Utah, and Andrew Alexander, W.M. Keck Laboratory for Functional Brain Imaging and Behavior, University of Wisconsin-Madison. Retrieved from http://en.wikipedia.org/wiki/File: DTI-sagittal-fibers.jpg. Reprinted under a Creative Commons Attribution License.

Part II

Landscape

5

Where does morality live, and when is it home?

Enemy soldiers have captured and terrorized your town. Now they are killing any survivors they encounter with orders to shoot on sight. You and your family, along with several neighbors, are hiding in the basement of an abandoned house. You hear a group of soldiers outside discussing whether to raid the house for valuables.

Your baby starts to cry. You cover his mouth to hush the sound, knowing that if you remove your hand and allow him to breathe, the soldiers will hear his screams and will come inside and kill everyone, including the baby. The only way to save yourself and the others is to smother your own child.

Would you kill your infant to save everyone else?

Now consider a different hypothetical dilemma. You are unlucky enough to be at the helm of a runaway trolley, fast approaching a fork in the tracks. You survey the scene and notice a group of five railway workers on the tracks to the left going about their work, evidently unaware of the trolley barreling toward them. On the tracks to the right is a single railway worker who is also perfectly oblivious to his potential fate.

If you do nothing, the trolley will follow the tracks to the left, causing the deaths of five people. The only way to avoid their deaths is to hit a switch on the dashboard that will shift the trolley's course to the right, causing the death of one workman.

Would you hit the switch in order to avoid four more deaths?

Most people with "normal" brains answer no to the first dilemma and yes to the second, despite the fact that both scenarios boil down to a choice between causing the death of one person and allowing the deaths of several. People with damage to a particular part of the brain, on the other hand—the ventromedial prefrontal cortex (VMPFC)—are much more likely to answer yes to both.[1] Why? What is going on in the brains of people with this kind of injury that explains their atypical decision to kill the baby? And just as provocatively, why do "normal" brains think that killing one person to save several others is the right thing to do in some cases, but not in others?

Inside the scanners, something quietly radical is happening: Our commonsense understanding of the way we make moral decisions is falling apart. Most of us go through life assuming that we know the difference between right and wrong, that our rational minds are capable of seeing those distinctions clearly and acting appropriately in light of them. We judge a woman more severely for killing her husband when we deem the act to be premeditated or "cold-blooded," but if she kills him in the "heat" of the moment, we are inclined to be lenient because we know that she was acting out of raw emotion, that she was somehow "not herself." We assume that unless a person is mentally ill or temporarily insane, she has the ability to make moral decisions rationally, "unhindered" by powerful emotions. Brain scans performed while we wrestle with moral dilemmas, however, are swiftly transforming our scientific understanding of how moral decision making works in the brain, and it turns out to look anything but "rational"—at least in our commonsense understanding of the word as the ability of the human intellect, as distinct from emotion, to apply a set of uniform, logical rules to moral decisions.

Knotty questions about morality, blame, and punishment provide abundant raw material for brain researchers who are designing experiments to determine, among other things, how brain scans of violent criminals differ from brain scans of "normal" people, and—on the flip side of the behavioral spectrum—whether "normal" brains are wired for altruism and cooperation. Do teenaged brains make moral decisions differently than adult brains, aging brains differently than younger adult brains? What are the implications of this deluge of new neurological information for our current system of blame and punishment? To the extent that we define and judge our "selves" by our behavior and our motives for it, the seemingly solid structure of the self is dissolving before our eyes.

Right or wrong, up close and personal

In 1992, prosecutors struck a deal with Herbert Weinstein, a 65-year-old advertising executive, after the judge allowed into evidence brain scans showing an abnormal cyst in Weinstein's arachnoid membrane (a delicate, web-like membrane surrounding the brain and spinal cord). He had been accused of murder after strangling his wife and throwing her body out the window, and the judge declared that lawyers for the defense could not tell jurors that arachnoid cysts were associated with any type of violence. Even so, on the morning of jury selection, prosecutors allowed Weinstein to plead guilty in exchange for a reduced charge of manslaughter, out of apparent fear of how a jury might react to seeing scans of his brain.[2]

Since Mr. Weinstein's case was up for trial, neuroscientific evidence has figured highly in a growing number of criminal cases, especially in the sentencing phase, and some experts in neuropsychology argue that new brain imaging work on our moral decision-making behavior points to a need to overhaul the American legal system. "New neuroscience will change the law," say Joshua Greene and Jonathan Cohen, psychologists at Harvard and Princeton, respectively, "not by undermining its current assumptions, but by transforming people's moral intuitions about free will and responsibility.... We foresee, and recommend, a shift away from punishment aimed at retribution in favor of a more progressive, consequentialist approach to the criminal law."[3] What do Greene and Cohen see coming out of the scanners, what "vivid new illustrations" of our strings (and what is or isn't pulling them) do they see bolstering arguments against retributive justice?

In September 2001, Greene, Cohen, and colleagues published a breakthrough study in which they scanned the brains of college undergraduates using fMRI while presenting them with a battery of 60 practical dilemmas. These dilemmas were selected from three distinct categories of problems. First were "nonmoral" dilemmas, like whether to travel by bus or by train given certain time constraints. Next were "moral-impersonal" dilemmas, like the trolley problem or like pocketing money from a lost wallet. Finally, there were "moral-personal" dilemmas, described as more "up close and personal," a category that investigators predicted would engage the brain's emotion circuits to a greater degree than the other two types of decisions.

The moral-personal category included a version of the classic footbridge dilemma, a scenario that has been juxtaposed with the trolley problem in moral philosophy classes since the 1990s[4] to tease apart how our moral intuitions often come into conflict. Imagine again a runaway trolley, except this time you are not a passenger but an observer, standing on a footbridge above the tracks. You see the trolley barreling toward five railway workers, who again seem oblivious to the trolley's approach. Standing next to you is an extremely large stranger whose body is clearly heavy enough to stop the oncoming trolley. Should you push him off the bridge and onto the tracks, sending him to a certain death in order to save the five workers? Most people say no, even though both the trolley and the footbridge dilemmas involve the choice of sacrificing one life for several.

"How is it," Greene and colleagues ask, "that nearly everyone manages to conclude that it is acceptable to sacrifice one life for five in the trolley dilemma but not in the footbridge dilemma, in spite of the fact that a satisfying justification for distinguishing between these two cases is remarkably difficult to find?" They hypothesize that "from a psychological point of view, the crucial difference between the trolley dilemma and the footbridge dilemma lies in the latter's tendency to engage people's emotions in a way that the former does not."[5]

To test this theory, they compared activity in areas of the brain associated with emotion and areas associated with working memory when the college students were tackling the three distinct types of problems. Prior imaging studies had shown that areas in charge of working memory become hushed when we process emotions as compared to when we do cognitive "work" that we find relatively unemotional.

The fMRI scanner recorded dramatic proof in support of their hypothesis: Areas of the brain associated with emotion dominated in the moral-personal dilemmas, whereas areas associated with working memory came on strong in both the moral-impersonal and nonmoral scenarios. In terms of the brain systems that were activated in each case, students processed the moral-impersonal problems in a way that more closely resembled nonmoral problems than moral-personal ones. In other words, their brains treated the trolley problem (whether to cause the death of one person to save several others) as if it were more akin to the choice between two coupons at the grocery store than to the stranger-on-the-footbridge problem (whether to cause the death of one person to save several others).

Greene and Cohen emphasize that the study was a descriptive and not a prescriptive one—that they were not attempting to prove any particular action or type of judgment to be right or wrong—but the implications of their findings for our understanding of morality run as deep as our sense of personal identity. Moral judgment in normal brains, says Greene, is "not a single thing; it's intuitive emotional responses and then cognitive responses that are duking it out."[6]

For people with VMPFC damage, on the other hand, rational utilitarian calculation (greatest good for the greatest number) apparently rules the day, "unhindered" by emotion.[7] The VMPFC is a region of the brain critical for the normal generation of emotions—in particular, social emotions—and the investigators concluded that "for a selective set of moral dilemmas, the VMPFC is critical for normal judgments of right and wrong," and that the findings "support a necessary role for emotion in the generation of those judgments."[8]

How will understanding the neural complexity beneath our seemingly straightforward moral choices change our attitudes about our moral intuitions? For starters, we no longer get to pretend that healthy brains are rational, emotionally sterile calculators spitting out morally consistent answers. Objective moral truth may or may not exist, but we can no longer assume that it is the only thing pushing our buttons.

The resultant shift in the way we view ourselves and our behavior may feel radical, but Greene argues that it need not feel apocalyptic. On the contrary, he finds the potential to redefine human morality liberating. "A world full of people who regard their moral convictions as reflections of personal values," he says, "rather than reflections of 'the objective moral truth' might be a happier and more peaceful place than the world we currently inhabit."[9]

Young brains, old choices

The capital case, *Roper v. Simmons*, showcased one of the most influential uses of neuroimaging evidence to date when the U.S. Supreme Court heard a challenge to a Missouri state law allowing the execution of juvenile criminals. Christopher Simmons, aged 17 and a junior in high school, had enlisted the help of two younger friends in the brutal murder of a woman, and he was tried, convicted, and sentenced to death for the crime in a Missouri court. The state supreme court later found the death penalty for juveniles to be unconstitutional, and the U.S. Supreme Court supported that conclusion two years later, basing its opinion largely on evidence of incomplete brain development in adolescents.

"There is little doubt that Simmons was the instigator of the crime," wrote Justice Kennedy in the majority opinion affirming that the law was, indeed, unconstitutional. "Before its commission, Simmons said he wanted to murder someone. In chilling, callous terms, he talked about his plan...Simmons proposed to commit burglary and murder by breaking and entering, tying up a victim, and throwing the victim off a bridge."[10]

The crime unfolded essentially according to plan, except that one of the boys backed out. (That boy was charged with conspiracy, a charge that was dropped in exchange for his testimony against Simmons.) The other two boys went ahead with the plan, breaking into the victim's home in the middle of the night, covering her eyes and mouth, binding her hands with duct tape, and driving her to a state park. On a railroad trestle overlooking the river, they tied her hands and feet together with electrical wire, wrapped her face in duct tape, and threw her off the bridge. Fishermen recovered her body the next day.

The vicious nature of the crime notwithstanding, the American Psychological Association (APA) and the American Medical Association (AMA) filed amicus briefs in the case challenging the death penalty for juveniles as cruel and unusual punishment, and therefore as unconstitutional under the Eighth Amendment. Both briefs relied heavily on neuroimaging evidence to argue that adolescents' behavioral immaturity is due, in large part, to underdevelopment of certain critical brain structures and over-reliance on others.

"First, adolescents rely for certain tasks, more than adults, on the amygdala, the area of the brain associated with primitive impulses of aggression, anger, and fear," the AMA brief maintained. "Adults, on the other hand, tend to process similar information through the frontal cortex, a cerebral area associated with impulse control and good judgment. Second, the regions of the brain associated with impulse control, risk assessment, and moral reasoning develop last, after late adolescence."[11] The brief cited imaging evidence that "the brain's frontal lobes are still structurally immature well into late adolescence," referring to the incomplete processes of myelination, the insulation of nerve fibers (this process speeds communication between cells), and "pruning," the natural thinning of the brain's gray matter (this improves the functioning of remaining neurons).

The Court was convinced. It agreed that adolescents' capacity for moral decision making is critically underdeveloped, and therefore that teens are less morally blameworthy than adults and less deserving of the ultimate punishment for their crimes. "The differences between juvenile and adult offenders are too marked and well understood to risk allowing a youthful person to receive the death penalty despite insufficient culpability," Justice Kennedy wrote in the majority opinion. "When a juvenile offender commits a heinous crime, the State can exact forfeiture of some of the most basic liberties, but the State cannot extinguish his life and his potential to attain a mature understanding of his own humanity."[12]

The Supreme Court finding drew a distinction between the "unfortunate yet transient immaturity" of youth and the "irreparable corruption" required for adults to commit similar crimes. The idea that at some point corruption becomes "irreparable" meshes with the popular assumption that by early adulthood the brain has lost its ability to change, and therefore that the worst offenders cannot be rehabilitated. As we've seen in a variety of areas, the notion that the adult brain is "hardwired" and immutable has been shown false time and time again, and may well be the next legal frontier for neuroimaging research.

Prevention versus punishment

A founding assumption of our legal system—that individuals can routinely make moral decisions without interference from emotion—is crumbling under the scrutiny of neuroscience, but it is the system's emphasis on retribution that Greene and Cohen ultimately wish to challenge. The commonly held moral intuition that criminals, unless they can prove mental defect, are bad people who should atone for their brazenly immoral choices depends upon the illusion of free will, they argue, and ultimately on the notion that the mind is separate from the brain—that a metaphysically distinct mind is the real decision maker operating the machinery of the brain, which in turn controls the body. In this view, the mind is responsible for its actions unless a malfunctioning brain intervenes, in which case the mind can't be blamed for its brain being broken, any more than a trolley driver should be blamed for damage done by a runaway trolley. "At this time," Greene and Cohen write, "the law deals firmly but mercifully with individuals whose behavior is obviously the product of forces that are ultimately beyond their control. Someday, the law may treat all convicted criminals this way. That is, humanely."[13]

Images of the brain in the throes of moral choice are unlikely to resolve debates over the mind-brain connection anytime soon, but they are playing an increasingly important role in our understanding of who might be at risk for violent behavior and what types of interventions (for example, drug therapy and/or talk therapy) might prove most effective for at-risk individuals. "The first important step," Richard Davidson and colleagues wrote in a 2000 review of the neuroscientific literature on violence, "is to recognize that impulsive aggression and violence, irrespective of the distal cause, reflect abnormalities in the emotion regulation circuitry of the brain."[14]

Some experts in forensic psychiatry say, so what? Neurological excuses are no different than sociological or cultural ones. Stephen J. Morse, professor of law and psychiatry at the University of Pennsylvania, puts it this way: "Some people are angry because they had bad mommies and daddies and others because their amygdalas are mucked up. The question is: When should anger be an excusing condition?"[15]

Neuroscientific evidence is beginning to weigh in on the issue of criminal responsibility for people with diminished mental capacity—people with brain damage, mental illness, or addictions. Should people with diseases or disabilities affecting the brain be held responsible for their criminal actions, and if so, when? "It's important for us to understand more about how brains work for people who aren't normal adults—juveniles, people with brain damage, people with mental disabilities from birth," says Henry T. Greely, director of Stanford University's Center for Law and the Biosciences. "They account for a large segment of our population, and a particularly large segment of our troubled population—of people who get into trouble *because* of their brains."[16] Understanding what is different about the brains of troubled youth and adults, Greely argues, will go a long way toward treating some of the systemic social ills that arise "because of these abnormal brains," and toward treating the underlying mental problems themselves.

Even if we settle questions of responsibility in obvious cases of diminished mental capacity, it still leaves the rest of us to consider—ostensibly normal adults, going through life helping and hurting each other, making moral decisions with our physically sound, healthy brains. How should we go about defining what counts as legally "normal"? What can pictures of healthy brains making moral decisions in different contexts teach us about how we should judge criminals in possession of apparently average brains?

"Since law is about human behavior," Morse says, "and we think the brain is at the foundation of human behavior—if we understand the brain and understand its relation to human behavior—this has potentially wide-reaching implications for law. Lots of people, for example, think that as we understand the causation of behavior, this will be a window into understanding whether people can be responsible for themselves."[17] He acknowledges that we are at the edge of a potential legal revolution, to the extent that a better understanding of the brain might make us think more mechanistically about ourselves and about "our view of responsibility, our view of the way we ought to interact with one another, our view of the rights we have, both against each other and against the state."

Morse, for one, predicts that our ideas about human responsibility will not be radically altered by images we see coming out of the scanners. "I think that understanding that there are causes for behavior, including neuroscientific causes for behavior, should surprise no one," he says. "I think we're far enough along in the understanding of how the world works to believe that this is a causal universe, and that we human creatures who are part of this causal universe are subject to the same causal laws as all the other phenomena of the universe." Morse argues that as far as we know, we're the only creatures to act for conscious reasons, the only creatures "who are capable of being responsive to reasons and of using reasons in our everyday lives. I don't think that there is anything that neuroscience is likely to find that is going to tell us that that picture of ourselves is radically false," he says.

For advocates of a justice system grounded in prevention rather than retribution, the crux of the matter is not whether neural activity is an excuse, but whether it is an explanation. If we understand the mental activity underwriting aggressive behavior, are we not in a better position to prevent future violence?

Our cultural fascination with the moral core—with how, exactly, humans go about discerning right from wrong (or not)—runs as deep as any question we ask about ourselves, and it provides unlimited fodder for our most popular dramatic narratives. In the acclaimed film *The Reader*, actress Kate Winslet deftly breaks our hearts as former SS guard Hannah Schmitz, convicted and sentenced to life for standing by while prisoners died in a fire—all the while insisting that she had no choice but to follow the rules. Ralph Fiennes, playing her former lover, asks her many years later if she ever thinks about the past, and if so, what she feels about it. "It doesn't matter what I feel," she replies. "It doesn't matter what I think. The dead are still dead."

Thanks to our ability to empathize, and Ms. Winslet's ability to act, we feel the simple truth of her words and her character's poignant longing for things to be different. Who among us hasn't wished that our remorse for past transgressions was all it took to reverse them?

As we move toward a clearer scientific picture of how moral choices get made, it seems that almost nothing matters more to achieving less harmful, more compassionate behavior (despite the theatrical declarations of Hannah Schmitz) than "what I feel"—the influence of our emotions on our moral decisions. From a violence-prevention standpoint, perhaps no human qualities are more vital than the mental dexterity it takes to work with powerful emotions, and the clear understanding that what we feel to be morally right is not only shaped by our values, but also colored, in any given instant, by our emotional experience.

Watching our judgments

However our moral choices get made—however deeply those choices engage our emotional circuitry—humans have been known, on occasion, to do bad things. Thanks either to too little empathy and compassion, or to too much anger and greed, we harm ourselves and others, and—we're the first to admit—we may or may not choose to do it again. Whether we believe in a justice system based on retribution or deterrence, the practical problems of assigning responsibility for wrongdoing, assessing criminals' potential for rehabilitation, and preventing future harms need to be addressed. As we refine our understanding of the neural circuitry involved in questions of right and wrong, how should that new information inform the nitty-gritty work of judges, jurors, and attorneys making life-and-death decisions in the courtroom? It is tempting, of course, to use every tool at society's disposal—including brain scans of the accused—to try to understand exactly what went wrong and why. Brain scans are among the most fascinating and irresistible pieces of physical evidence available, so it is no surprise that they are being introduced into courtroom proceedings at an unprecedented rate. What are the benefits and potential dangers of using information produced by this new—sometimes uncertain—science to determine the fate of the accused? When can interpretations of brain images be considered reliable and informative, and when are they simply too speculative or prejudicial to serve the ends of justice?

Neuroscientist Michael Gazzaniga, director of the SAGE Center for the Study of Mind at the University of California, Santa Barbara, thinks there is no time like the present to deal systematically with questions about the proper use of neuroscientific evidence in legal contexts. He has spearheaded the Law and Neuroscience Project, a ten-million-dollar effort funded by the MacArthur Foundation, to begin to sort out these issues at a national level. The project pulls together experts from all over the country in the fields of neuro-science, psychology, law, and philosophy—including Joshua Greene, Henry Greely, and Stephen Morse—to debate the pros and cons of brain scans as evidence and how new information about the brain should and should not reshape our system of blame and punishment.

"We're doing this project now," Gazzaniga says, "because in recent years, there's been an explosion of information about human neurobiology—as opposed to the neurobiology of animals—and this explosion has raised new questions, profound questions, about what we can know about the human mind and about how that information can be used and misused."[18] In the context of the judicial system, neu-roscientific evidence might be used quite appropriately to assess the mental status of defendants and witnesses, but it might also be mis-used by people wishing to overstate the current value of brain scans to detect lies or to uncover malevolent intentions.

Whether neuroimaging evidence holds the potential to read the minds of defendants, accusers, and witnesses "remains very much an open question," Gazzaniga says. "Studies that are underway may well be able to reveal whether someone has actually had a particular expe-rience in the past, even though they may deny it."[19] The technology is not there yet, he says, "but it most likely will occur at some point in the next 10 or 20 years." In the meantime, one of the functions of the project will be to examine "both sides—where [neuroscientific evi-dence] is used properly and where it's misused."[20]

Walter Sinnott-Armstrong, professor of philosophy at Dartmouth College and codirector of the project, stresses the extreme urgency of evaluating these problems of evidence because information from scanners is finding its way into an increasing number of new criminal cases. "Right now, neuroscience is entering the courtroom,"[21] he says. "Lawyers are using it for their clients' benefit in various ways. Com-panies are developing methods of lie detection. And we have to, to a certain extent, put the brakes on and make sure that it doesn't get abused. If we don't step in and have an impartial body look at those uses of neuroscience in the courtroom, and decide whether they really are reliable enough for that context, then we're going to have a lot of trouble down the line."

Initially, project members are tackling issues of criminal responsibility and how neuroscientific research should—and should not—inform the judicial process. "Surely the neuroscience here is not magic,"[22] says Owen D. Jones, professor of law and biological sciences at Vanderbilt University. "It doesn't enable us to read minds. But, we hope within that limitation, to understand the context in which it might be used effectively and no further. So we hope to help legal actors—judges, jurors, legislators—have a more developed sense of where the promise is in the technology and also where the limits are." Jones's own research has focused on the brain activity underpinning our decisions about whether or not a person is guilty—and if so, just how severely to punish them—and he has been appointed codirector of the project's working group on decision making and the law. His findings dovetail with Joshua Greene's on moral decision making in that they show that when people are asked to dish out punishment for a variety of hypothetical criminal scenarios, the degree of activity in emotional circuits in the brain predicts the severity of their recommended penalties.

Advocates of neuroimaging evidence tend to compare its factual value to other, more established forms of physical evidence such as DNA, but critics say the comparison is misleading. "The reason DNA is so powerful," Gazzaniga notes, "is because it is virtually the raw truth."[23] Neuroscientific evidence, on the other hand, is "a long way from that kind of certainty," because it tends to be more correlative in nature. A particular pattern on a brain scan can be associated with a particular thought or emotion—dysregulated anger, for example—but is that one state of mind the only explanation for a particular snapshot of an individual's brain? "When you have a particular biologic state that you have identified with new brain imaging," Gazzaniga says, "does that necessarily mean that person is holding a particular thought or has a certain intention or not?" He suspects that future brain research might nail down many of these issues, but for the moment, he says, we're "not there yet by a long shot." Even if we get there, we will need to address the issue of whether brain scans fall under the protection of the Fifth Amendment—whether they count as verbal self-incrimination—and this question will probably warrant at least one high-profile Supreme Court case.

Henry Greely is codirector of the project's working group on diminished brains and the law, and he stresses that neuroimaging research is as fallible as any new science and that we need to proceed with caution while we learn to use it accurately and responsibly in a legal context. "Neuroscience is a double-edged sword for society," Greely says. "On the one hand, it will be able to tell us many things that will be important, and we have to adjust, we have to prepare for those. On the other hand, it's a new science. A lot of the findings are going to turn out to be wrong—not because they're bad scientists, but because that's the way science works. You have huge flowering at the beginning, and then you prune it down as you understand better what's true and what's not true."[24] One of the real threats to society, he says, is that we might be much too hasty to seize on some of these findings as right, and "to go down what turns out to be a very bad path. I think one of the most important things this project can do is help sort out which paths are promising, which paths aren't—which solutions work, which solutions don't—because keeping us from going down the wrong path is just as important as helping us go down the right path."

Gazzaniga delineates the current limitations of neuroscientific evidence quite bluntly when he says, "This is baby science, first-step science, like genetics in the 1950s. This should be used cautiously in the courtroom—if at all."[25] Morse is similarly wary of relying too heavily on a nascent science to tip the scale of justice toward guilt or innocence. "What tends to happen with science is that people get starry-eyed about what it can do for the law,"[26] he says. "Then the train leaves the station heading in the wrong direction."

At least as fascinating as the issue of how brain scans should be *offered* as evidence in courtrooms is the issue of how brain scans should be used to better understand how evidence is *evaluated* in courtrooms. "Courtrooms are just filled," Sinnott-Armstrong says, "with moral judgments. When a jury or judge decides that a certain person deserves a certain penalty—a certain sentence in prison or a certain type of treatment—they're making a moral judgment. And very often, even implicit moral judgments that they're not aware of will affect the kinds of decisions that are made by the legal system."[27]

To make the courtroom process work the way we hope it will—fairly and impartially—the project is looking at how people reach conclusions about the moral decisions of others. These conclusions are "crucial for the courtroom process," Sinnott-Armstrong says, "and so we need to understand the basis on which people are making them, in order to try to make the process work better."

Owen Jones's research demonstrates that supposedly rational, impartial decisions by judges and juries are anything but unemotional. "We take decision making for granted, like breathing," he says. "If you want a world in which judicial and jury decisions are fair, unbiased, sensible, and reasonable, then we ought to understand a little bit about how it actually happens."[28]

He and his colleagues at Vanderbilt analyzed brain scans of 16 volunteers when they were asked to make decisions about guilt and punishment for a range of hypothetical crimes, from petty theft to rape and murder. Participants were asked to read stories that described scenarios involving a character named John, stories that fell into three categories: "responsibility," "diminished responsibility," and "no crime."[29] In the "responsibility" scenarios, John intentionally commits a crime and is considered fully accountable for his harmful actions, whereas in the "diminished responsibility" scenarios, he commits similar crimes but is considered less responsible due to mitigating circumstances. In one such scenario, for example, John burglarizes a home under orders from a violent drug dealer who has threatened to harm his daughter if he refuses; in another, John becomes increasingly volatile while suffering from a brain tumor and ultimately abducts, tortures, and kills a small child. In the no-crime scenarios, John's actions might have unforeseen harmful consequences, but his behavior is not considered criminal.

After reading the scenarios, participants were asked to rate them on a scale from 0–9, according to how much punishment they thought John deserved for his actions. The fMRI scans showed that one area of the forebrain—the right dorsolateral prefrontal cortex (rDLPFC)—became active when people distinguished between scenarios on the basis of responsibility. It was engagement of circuits linked to social and affective processing, however, that predicted the severity of punishment; the more intense the level of activation in several emotional centers of the brain—including the amygdala—the more dearly participants thought John should pay for his actions.

One of Jones's partners on the research, Vanderbilt neuroscientist Rene Marois, says he was shocked by the degree to which emotional circuits are involved in so-called impartial legal decisions. "This reasoning may not be so detached," he says. "It shattered my preconceived ideas of the legal system. But for a lawyer, maybe it doesn't."[30]

In the upcoming five years or so, Gazzaniga says that we're going to develop "quite a rich picture of what we come with 'from the factory,' as it were, and how much of it is learned from the local culture we're in."[31] He believes that these findings about what neural patterns we are born with and what is culturally determined will have a profound effect on how we think of ourselves as humans and how we choose to judge fellow members of our species.

"I firmly believe that this century is the century of the brain," Greely says. "We are learning so much more about how our brains work, and our brains are the essence of our humanity. When we interact with other people, we may think we're interacting with their bodies, but what we really care about is our interactions with their brains. Our society, I think, is a society of human brains much more than it is of human bodies."[32] We are, Greely believes, in the process of understanding vastly more about how the brain works in social and moral contexts than we ever did before. "That will change our society," he says. "We need to be ready for that."

Wired for selflessness

In a world where entire communities have been left devastated by war and violent crime, the "baby in the basement" dilemma is more than a mere mind experiment. That humans possess natural tendencies toward competition and, under the wrong conditions, even extreme violence, seems uncontroversial. Yet wherever we turn, we see ourselves being awfully good to one another. Examples of selfless behavior abound, from the minor to the extreme—from shoppers sacrificing their place in the grocery line, and families donating cans and jars to food banks even when their own budgets are squeezed tighter than ever; to soldiers taking enemy fire to protect comrades, and civilians jumping into freezing rivers and charging into burning buildings to save the lives of perfect strangers. Our most dramatic acts of altruism seem to happen when we don't have time to reflect because we simply have to do. This is when altruism seems absolutely instinctual—perfectly devoid of any selfish motive—which raises the question of who or what is pulling our strings when we sacrifice our own needs and desires, our own safety, even our own lives for the safety and happiness of others.

Evolutionary biologists have long theorized that tribes of early humans needed cooperative skills to thrive in groups and therefore to survive as individuals—an explanation that might work just fine to explain our kind behavior at the grocery store or the food bank, but that comes up lacking when we're plunging into an icy river. The evolutionary origins of some of our most spectacular displays of altruism may always elude the lens of science, but cognitive scientists pose equally intriguing questions about our present-day motives for giving. Using real-time fMRI imaging, we can watch brain activity that coincides with self-sacrificing and cooperative behavior, and we are learning that our altruistic tendencies are just as much a part of our wiring, just as automatic and "natural" a part of our brain activity, as violence-provoking emotions like fear, lust, and anger.

Psychologist Ulrich Mayr and economists William Harbaugh and Daniel Burghart of the University of Oregon wondered what it is exactly that "motivates people to see beyond themselves."[33] They designed an fMRI study to tease apart the motives of "pure altruism," the satisfaction we take from increases in the public good regardless of our own involvement, and what has been called the "warm glow" effect, the pleasure we take in playing the role of benefactor. Some economists have argued that it is always the warm glow effect that motivates our altruistic behavior—that there is no such thing as pure altruism when it comes to charitable giving—but the Oregon study indicated otherwise. FMRI images showed that donating money to a local food bank, and to a lesser degree, paying an involuntary "tax" that would benefit the food bank, activated the brain's reward centers—the same areas that brighten when we eat yummy desserts, get paid, or take recreational drugs. "The most surprising result is that these basic pleasure centers in the brain don't respond only to what's good for yourself," said Mayr. "They also seem to be tracking what's good for other people, and this occurs even when the subjects don't have a say in what happens."[34]

To look for purely altruistic and warm-glow motives, 19 female students were given a starting amount of $100 and were told that they would make a series of anonymous choices about how much of it (if any) to give to a local food bank and how much to keep for themselves. In order to eliminate any influence exerted by the fear of appearing greedy or the desire to appear generous, students were informed that investigators would not be aware of their choices—that these choices would be recorded inside the scanner on a portable memory device, and that even the lab assistants in charge of paying the balances would remain in the dark about who had given how much (because some transactions would not be voluntary). The students' brains were then scanned while they were presented with transactions—some voluntary and some forced (to resemble taxation)—and while they saw those transactions affect their accounts.

The nucleus accumbens and the caudate, areas of the brain associated with pleasure and reward as well as with learning and memory, and the insula, a region key to pleasure and to critical human emotions like empathy and compassion, lit up to similar levels of brightness when participants gave freely and when they received money themselves. Just as notably, these areas were also activated, albeit less intensely, when the giving was forced. "This is the first evidence we know of," the investigators wrote, "demonstrating that mandatory taxation for a good cause can produce activation in specific brain areas that have been tied to concrete, individualistic rewards."[35]

The pure altruism model of giving predicts that people who value seeing a charity benefit from a donation (regardless of whether they themselves were a voluntary benefactor)—and who value that benefit more than they value receiving benefit for themselves—are more likely to give. The investigators tested this model by dividing respondents into altruists and egoists (10 and 9 women, respectively), depending on whether their neural response was greater for the food bank's payoff or for their own. Sure enough, the altruists, those students who got a bigger neural bang for the food bank's buck, ultimately chose to donate about twice as much—$20, on average, as compared to the egoists' $11. "The larger a person's neural response to increases in the public good," the investigators noted, "no matter what the source, the more likely they will give voluntarily."[36]

This finding is provocative, in part, because it points to the potential "trainability" of social responsibility. "Many psychologists," Mayr and Harbaugh explain, "believe that the main evolutionary purpose of the 'pleasure areas' that we tracked in our experiment is reinforcement learning. Actions that activate the pleasure areas are more likely to be repeated in the future."[37] Now that researchers know that these same areas are engaged during altruistic behavior, we might next explore whether altruism is as teachable as other rewarding behaviors. Recent discoveries about the flexibility of neural pathways involved in emotion, attention, memory, and recovery from chronic pain and addiction are certainly suggestive of our potential to strengthen our "giving" circuits as well.

This trainability, if it bears fruit, might address a central question posed by Greene, Cohen, and colleagues: "How will a better understanding of the mechanisms that give rise to our moral judgments alter our attitudes toward the moral judgments we make?"[38] In his article "From Neural 'Is' to Moral 'Ought,'" Greene highlights a pair of moral dilemmas with obvious applications to modern life (adapted from writings of philosopher Peter Unger[39]). Imagine that you are driving along the road and you hear a desperate cry coming from the bushes by the roadside. You see a man lying on the grass, his leg covered in blood. He was hiking and had a bad fall, he tells you, and your initial inclination is to get him to a hospital immediately to save his leg. You also realize that the blood gushing from his injury will ruin the leather seats in your new car. Is it acceptable for you to drive on without helping the man to save your expensive upholstery? Most people say absolutely not; that would be monstrous.

Now imagine yourself sitting at home, reading the mail. You have received a letter from a well-respected international aid organization asking you to make a $200 donation that would provide life-saving medical care to people in a poor country who otherwise would be unable to afford it. Is it morally acceptable not to make the donation to save yourself $200? Most people say yes; in this case, it would be morally acceptable not to spend the money, even though both cases involve the opportunity to provide lifesaving medical care to others at a relatively modest material cost to oneself. Why are our intuitions here so different?

"Maybe there is 'some good reason' for why it is okay to spend money on sushi and power windows while millions who could be saved die of hunger and treatable illnesses," writes Greene. "But maybe this pair of moral intuitions has nothing to do with 'some good reason' and everything to do with the way our brains happen to be built."[40] Greene hypothesizes that we may have evolved to weigh up-close-and-personal moral decisions differently than physically distant ones, not because our moral decision making reflects a reasoned understanding of our moral obligations, but because our ancestors evolved in tightly knit communities in which the only lives we could save were the ones we interacted with face-to-face. Decisions about whether to send money to Oxfam, on the other hand, might not engage our primal emotional circuits to the same degree because "our ancestors did not evolve in an environment in which total strangers on opposite sides of the world could save each others' lives by making relatively modest material sacrifices."[41]

If neural circuits that support selfless giving do indeed turn out to be flexible and trainable with practice, perhaps we won't need to depend on the slow grind of evolution to shape our brains in response to modern moral dilemmas—potentially our saving grace in a world where humanitarian and ecological crises routinely outpace our ability to address them.

6

The making and breaking of memories

Can't find your glasses? Temporarily mislaid the children? Don't despair—"Reclaim your brain."[1] This is the bold promise of Lumosity.com, a website offering brain-training exercises for a monthly fee, for those of us whose gears are cranking a bit slower these days—perhaps even grinding to an embarrassing halt altogether. There are video games, too, marketed to a swelling audience of boomers attracted by any and all products that might act as antidotes for neuronal loss (applied sparingly but regularly—like Rogaine for the brain). Nintendo's Brain Age claims to "Give your brain the workout it needs." The website pictures a happy middle-aged mom and dad on the couch with their teenaged daughter, and all of them are wielding their styluses to perform handheld mental calisthenics. "Just minutes a day, that's all it takes to challenge your mind."[2]

Can we trust that there is solid scientific evidence behind these confident commercial claims? Are they supported by the latest brain-imaging research? If so, what's to become of those of us with no expendable income to enjoy luxuries like Nintendo or online subscriptions? Is there any hope for our economically disadvantaged neurons?

All in all, the news looks excellent for those of us—financially secure or otherwise—who've made it past 30 and who feel that we are, quite frankly, losing it. (Whatever "it" is—we honestly can't remember, but we know we used to have it.) The absence of "it" might be as innocuous as regularly mislaying our car keys or forgetting why on earth we walked into the kitchen. Or it might be as irksome as constantly forgetting the name of, um, our neighbor what's-her-name, or discovering that we hung up the phone in the fridge again. It might even turn out to be as dangerous, or as frightening, as forgetting the way home, or whether we happen to love the people living there.

Fear of the cognitive worst is driving more of us to challenge our brains the way we have learned to challenge the rest of our bodies to stay strong and healthy well into our golden years, and all indications are that it's a smart idea that will only make us smarter. Neuroimaging research provides the foundation for reputable brain-training programs, of which Lumosity is one; designed by brain scientists at Stanford and UCSF, this program has been shown effective at enhancing memory and cognitive agility in a clinical study. Cognitive training programs will likely play an increasing role in routine preventative care, and meanwhile, imaging research is pointing in another, less obvious direction: the gym. Even light physical exercise enhances mental functioning and slows the effects of age-related cognitive decline, researchers tell us, and for those of us feeling the neural wear and tear of the years, this is heartening news. Put the right tools in our hands; we will do the necessary repairs.

These hopeful findings have to do with memories we'd like to keep. What about traumatic or destructive memories we'd much rather forget? Some of us go through life feeling held hostage by intrusive memories, feeling as if our lives are excessively defined by the fear and dread they evoke. Emotionally charged memories can determine the types of risks we decide to take and the people we choose to know; looking back, we might feel as if they've directed the entire course of our lives. In extreme cases, living with traumatic memories can decimate our physical and emotional health.

Imaging research is confirming what cognitive scientists have suspected for years: memory and emotion are intimately linked in the brain, and altering our experience of one can drastically affect our experience of the other. As neuroscientists learn how emotionally charged memories are encoded, they are also learning how to soften the effects of painful ones. There are even new indications that we might be able to erase them altogether, much as we would an old VCR tape we no longer care to watch. As we develop the ability to rid ourselves of the effects of unwanted memories, will we chip away at the core of what we think of as "me?" Are painful memories an integral part of our personal identity, to be retained at any cost, or are they a mere biochemical configuration that can, and should, be altered or even expunged in the interest of preserving our emotional and physical health? Though people disagree wildly on these issues, our curiosity continues to uncover the anatomical particulars of memory with breathtaking speed.

Holding onto history

Alzheimer's disease (AD) is the most common form of dementia, affecting as many as 4.5 million Americans and more than 26 million people worldwide. These numbers could quadruple by 2050, as rates of the disease keep pace with an exploding population and longer life expectancies.[3] Alzheimer's is not a part of normal aging, but the risk of the disorder does increase with age. About 5 percent of people between the ages of 65 and 74 have AD, whereas nearly half of people over the age of 85 live with the disease. The devastating illness is characterized by the formation of abnormal clumps of protein (amyloid "plaques") and twisted bundles of fibers ("tangles") in the brain, and by the resulting loss of healthy neurons—anatomical changes that can take hold 10 to 20 years before symptoms emerge and eventually lead to memory loss, mental and social impairment, personality changes, and ultimately death.

Age-related cognitive decline, by contrast, is something we all expect to contend with, and we are usually forced to sooner than we hope. Recent studies suggest that for some of us, a decline in certain mental abilities—puzzle solving and abstract reasoning, for example—might appear as early as our mid-20s.[4] Prevention, diagnosis, and early treatment of memory problems—both of the "normal" aging and Alzheimer's types—are among the hottest priorities in neuroscientific research today, and thanks to new imaging tools, many of memory's most salient secrets have been exposed in the last five years.

In September 2004, neuroscientist Randy Buckner published a massive review of imaging research in the journal *Neuron* demonstrating beyond a doubt that Alzheimer's disease attacks different parts of the brain than those affected by normal aging. "We're getting a better understanding of the complex constellation of factors that change [in the brain] with aging," Buckner said. "When you start to look across the literature, lots of data points converge, suggesting there are certain changes that take place in aging that are not what cause Alzheimer's disease."[5] Buckner highlighted changes in frontal regions that occur in normal aging, including in the corpus callosum, the tract of white matter that serves as the phone line between the two cerebral hemispheres. These brain changes were associated with decreased executive function—the ability to complete complex problem-solving or goal-oriented tasks. (Some candidate causes are hypertension, which can lead to vascular damage, and depletion of neurotransmitters with age.) Long-term memory declines seen in Alzheimer's disease, on the other hand, were associated with changes in the medial temporal lobe system, including the hippocampus, the seahorse-shaped structure deep in the forebrain with major responsibilities for processing memories and emotions.

All these changes are strongly associated with age, but they seem to affect memory quite differently across individuals. "An important further factor in cognitive aging is how an individual responds to change," Buckner wrote in his review. "Growing evidence suggests that compensation for brain decline in aging may partly account for why some older adults age gracefully and others decline rapidly."[6]

Buckner's earlier work had shown that—contrary to popular belief—some forms of memory are left intact by Alzheimer's. In a 2004 study, Buckner and colleague Cindy Lustig performed fMRI scans while giving a word classification task to 34 young adults, 33 healthy older adults, and 24 older adults with early-stage Alzheimer's. This sort of classification task involves implicit (procedural) memory as opposed to explicit memory (which is used to recall people, places, and events). All three groups showed improvement with practice, and the amount of time it took them to classify a word grew shorter—"the hallmark of implicit learning," Buckner said. Areas of the frontal cortex associated with high-level cognition and planning were most active in response to the word task, suggesting that despite damage to these areas in AD, some memory processes that rely on them remain relatively unaffected. "It appears that a number of brain systems are more intact in Alzheimer's than we had anticipated," said Buckner. "The findings suggest that if we can help people use these brain systems optimally by providing the right kinds of cues or task instructions, we may be able to improve their function."[7]

Buckner and colleagues have also produced exciting real-time fMRI images of how mental training can improve normal age-related memory decline. In a study involving 62 healthy people, some in their 20s and others in their 70s and 80s, subjects were asked to remember words while inside the fMRI machine. "For a long time, we have known that as people age they start to have difficulties with higher-level-controlled cognitive processes," Buckner said. "For example, sometimes older adults have difficulty in novel situations where they must respond flexibly to memorize things."[8]

Then the researchers conducted the experiment again, this time providing older adults with a memorization strategy. Instead of asking them simply to remember words, they presented them with one word at a time and asked them to categorize it—for example, as abstract or concrete. After working with the words in this way, not only did activity increase in high-level frontal regions of their brains, but their performance on the memorization task also improved. "The situation with regard to overcoming memory deficits in aging is much more promising than we thought," said Buckner. "It could have been the case that the frontal regions in the older adults had atrophied or undergone cellular deterioration to the extent that they were inaccessible to these individuals. But that was not the case."[9]

This is welcome news for us all, but what are the practical implications? How can we use this new information to improve quality of life for those of us in our elder years? "Our hope is that by demonstrating the availability of these [memory] systems," said Buckner, "this knowledge will be translated into cognitive training programs for the healthy elderly and those with forms of dementia, which we just had not anticipated when we began this work."[10] Science-based cognitive training programs might become as much a part of routine preventative care for our aging minds as dietary guidelines and medicines have become for conditions like heart disease and diabetes.

The dark network sheds light

Perhaps the most provocative idea coming out of recent memory research is the novel theory that a particular pattern of brain activity in healthy young people—a pattern of thought associated with day-dreaming, musing, or reminiscing about past events—might actually presage Alzheimer's disease. The gist of the hypothesis is that this type of thinking—dubbed "mental time travel" by some cognitive scientists—manifests itself in the same brain regions most devastated by Alzheimer's, and thus that it might contribute to a series of metabolic changes that lead to the disease. The default network is so named because it comes online when it appears that we are resting—when our brains are not engaged in externally oriented or goal-directed "work"—and it encompasses areas in the frontal, parietal, and medial temporal lobes. These regions of the brain are also referred to as the "dark network" because they remain dim when we're mentally connected to the moment and to our external world—when we are engaged in a task, solving a problem, or interacting with other people.

"The regions of the brain we tend to use in our default state when we are young are very similar to the regions where plaques form in older people with Alzheimer's disease," says Buckner, who led a 2005 analysis of imaging data collected on 764 people—some healthy, and others with varying degrees of Alzheimer's-related dementia. "This is quite a remarkable convergence that we did not expect."[11]

When young people are asked to focus on a task such as solving a math problem, reading a book, or actively listening to another person, their brains tend to switch smoothly from the default state to goal-oriented neural activity. When a patient diagnosed with Alzheimer's, on the other hand, is asked to concentrate on a task, the default network does not quiet down at all, but instead gets noisier, indicating a severe disruption of the normal balance of power between this network and areas of the brain devoted to goal-directed activity. (As we've seen, over-reliance on areas within the default network is a hallmark of another debilitating neurological condition, chronic pain.)

"We are very interested in exploring these new observations to understand who is at risk and who is protected from Alzheimer's," says Buckner. To develop effective treatments, scientists will need to catch the devastating disease in its earliest stages, when drugs and other forms of therapy might be most helpful in slowing its progress. "You have to get to this pathology before it has its biggest effect, before it has done its damage,"[12] says psychiatrist William Klunk, coauthor of the paper and head of the University of Pittsburgh team that developed a powerful tool for imaging Alzheimer's, a substance known as Pittsburgh Compound B, or PiB. This radioactive dye can attach itself to plaque deposits in the brain, allowing scientists to use PET scans to spot the disease in patients long before symptoms occur. "PiB is being used today to help determine whether drugs that are meant to prevent or remove amyloid [plaques] from the brain are working," Klunk says, "so we can find drugs that prevent the underlying pathology of the disease."[13] Besides monitoring the effectiveness of treatment, PiB might allow for earlier diagnosis and might distinguish Alzheimer's from other forms of dementia, helping doctors tailor the most effective treatment and management regimens for their patients.

Exercise the body, exercise the brain

Those of us concerned about sharpening our minds might not think to take a brisk walk, go for a swim, or chase after the kids, but regular physical activity—even light exercise—is emerging as one of the most powerful tools for the prevention of dementia and of "normal" age-related cognitive decline. A 2009 fMRI study out of Columbia University Medical Center revealed a major underlying mechanism of this phenomenon when it linked blood sugar control to activity in the dentate gyrus, a small area nestled in the hippocampus and thought to be key to the formation of new memories. Researchers compared the brain scans of 240 elders and found a strong correlation between high blood glucose levels and reduced blood flow to this vital memory center. The study looked at other measures that typically change with age, such as body mass index, cholesterol, and insulin levels, but only blood glucose levels correlated with decreased activity in the dentate gyrus.

"If we conclude this is underlying normal age-related cognitive decline, then it affects all of us," says Dr. Scott Small, lead investigator on the study and associate professor of neurology. Our ability to regulate glucose starts to decline in the third or fourth decade of life, Small notes. "Since glucose regulation is improved with physical activity, we have a behavioral recommendation—physical exercise."[14]

The correlation holds true when blood sugar levels are only slightly elevated—when they would not be considered indicative of disease states like diabetes or metabolic syndrome. "We're noticing subtle changes, and that is a valid reason to worry,"[15] Small says. But there are simple, sweaty steps we can all take to protect our cognitive skills. "By improving glucose metabolism, physical exercise also reduces blood glucose. It is therefore possible that the cognitive enhancing effects of physical exercise are mediated, at least in part, by the beneficial effect of lower glucose on the dentate gyrus. Whether with physical exercise, diet or through the development of potential pharmacological interventions, our research suggests that improving glucose metabolism could be a clinically viable approach for improving the cognitive slide that occurs in many of us as we age,"[16] Small concludes.

Other studies have linked diabetes and high blood glucose to memory deficits and dementia—and physical exercise to protective effects against these unwelcome changes. Small's research is a major key to the glucose puzzle. Exercise physiologist Sheri Colberg-Ochs, who has published widely on the relationship between diabetes and exercise, has demonstrated that even light physical exertion can protect against the erosion of cognitive abilities in patients with type 2 diabetes. She says that Small's research is illuminating because "it allows for a greater understanding of which region of the hippocampus is likely most affected by poorly controlled diabetes."[17]

Why, then, does elevated blood sugar have such a profound effect on this tiny but vital area of the brain? "The simple answer," says Small, "is we don't know yet."[18] The research to date shows that the dentate gyrus is the one area of the hippocampus that seems to be particularly sensitive to *any* fluctuations in glucose, he says, "whether it's up or down." (He notes that prior research has suggested that this brain region is also especially sensitive to abnormally low blood glucose.) These findings are critical, he says, because to understand how different factors accelerate cognitive decline, it is necessary to "really pinpoint the areas that are what we would call 'differentially vulnerable,' or especially sensitive to a particular process, whether it's a disease process—or in this case, glucose. It really frames the question. It's an important first step toward getting at the mechanism." He and fellow researchers are digging deeper for that mechanism, and any answers they yield might lead to targeted new treatments and management strategies for age-related memory decline.

Small has his own pet theories about the source of the trouble—theories that he emphasizes are "complete speculation." (He tells his students, "One should never be afraid of speculation. That's the fun part of what we do.") The dentate gyrus is unique within the brain for a number of reasons, not the least of which is its ability to generate new neurons—a process known as neurogenesis. Scientists used to believe that brain cells did not divide after birth. That dogma still seems to hold true for most areas of the brain, but it has been disproved for a couple of tiny neural regions including the dentate gyrus, which does support cell division after birth. "That's enormous news in the world of neuroscience," Small says, "and people are trying to understand, well, what does it really mean? What regulates this process?"

Why adult neurogenesis happens isn't clear, although many researchers hypothesize that it occurs in the hippocampus because the generation of new nerve cells is critical to learning and memory. How exactly learning and memory might be affected by this process is a mystery, but several theories are in the running. New neurons might increase memory capacity; they might reduce interference between memories; or they might add information to already existing memories (or all of the above).

One favorite hypothesis about the importance of lifelong neurogenesis is that neurons generated in adulthood might play a critical role in the regulation of stress. Studies have linked formation of new neurons to the beneficial actions of antidepressants, suggesting a connection between deficient hippocampal neurogenesis and clinical depression. This idea is consistent with findings that associate stress-relieving activities such as learning, exposure to a new (albeit unthreatening) environment, and exercise to increased rates of neurogenesis. It is also consistent with observations that animals exposed to physiological stressors (such as increased levels of cortisol) or to psychological stressors (such as intensified levels of isolation) show marked decreases in production of adult-born neurons.

The dentate gyrus is a key site for adult neurogenesis, and Small suspects that the hypersensitivity of this area to glucose throughout our lifetimes may be left over from the fact that it is designed to support new cell division after birth. "A dividing cell needs a lot more energy than a nondividing cell," he says, and so the dentate gyrus must be equipped with all sorts of molecules for regulating energy metabolism—"everything to do with glucose," he says. Each region of the brain has its own unique molecular profile; it expresses different genes. "No one really understands why, but this is probably one of the things that the dentate gyrus expresses—genes that regulate glucose—and that's why we suspect it's sensitive to any fluctuations in glucose," even later in life when it might turn out that there is no widespread production of new cells.

These findings are not just important for the eldest among us, researchers point out, but for the youngest as well, as American children and teens live at ever-increasing risk for type 2 diabetes. "When we think about diabetes, we think about heart disease and all the consequences for the rest of the body, but we usually don't think about the brain," Dr. Bruce McEwen, head of the neuroendocrinology lab at Rockefeller University, told the *New York Times* in January 2009. "This is something we've got to be really worried about. We need to think about [young people's] ultimate risks not only for cardiovascular disease and metabolic disorders, but also about their cognitive skills, and whether they will be able to keep up with the demands of education and a fast-paced complex society. That's the part that scares the heck out of me."[19]

Small agrees that abnormally elevated blood glucose is probably harmful to developing brains, but the good news for patients—young and old—living with diabetes is that "if you control your diabetes and you control your glucose—then presumably you're not affected," he says. "It's like high blood pressure. People have hypertension, but if they take their meds, their blood pressure is regulated."[20] Likewise, if people with diabetes and hyperglycemia can regulate their blood sugar, it will diminish the risk of cognitive problems down the line.

In light of revelations that high blood sugar can damage the aging brain, why do researchers emphasize physical exercise over dietary choices as a means of alleviating the problem? Why not simply advocate a particular type of diet to protect the brain from high glucose— a low-carb diet, for example?

"Diet is a tough one," Small explains, "because obviously, it's not just staying away from sugars. That won't work." Diet is an essential part of regulating conditions like diabetes and obesity, he says, but for anyone who still eats—including generally healthy individuals—there is no avoiding regular ups and downs in blood sugar. "The issue is, when you eat any meal—even low carbs—your body has a spike in blood glucose. Many foods are broken down into glucose, and of course you need glucose, so it would be a bad idea to eat nothing that provides it," Small says. The real issue, then, is how the body handles those spikes in blood sugar, and unfortunately our body's ability to process sugar ebbs as the years roll on. But no matter what our age, exercise is a well-established answer to the dreaded "sugar spike."

"Glucose absorption seems to be affected by aging," Small says. "So what you really need to get at mechanistically is the body's ability to sponge up glucose after a meal, and that's exactly what exercise does." His team had already found that physical exercise has a differential benefit in the dentate gyrus, but at the time, "we didn't quite understand. Then we made this glucose link, and it immediately provided a mechanism."

One of the many established benefits of physical exercise, he says, is that it improves the body's ability to handle blood sugar by inserting glucose transporters into muscles—so in a sense, "it allows your muscles to be spongier to glucose." People who exercise, he explains, "sop up glucose more quickly," and therefore they regulate it better. It's not just our intake of glucose, then, but our body's ability to absorb it efficiently that is critical to maintaining our mental dexterity well into our later years. Physical exercise, he says, is "something I recommend quite regularly to all my patients—and to myself, and to friends."

New animal research shows that voluntary physical activity helps mice learn faster and better and helps older mice generate new neurons in the hippocampus. The connections between exercise-mediated neurogenesis and learning remain uncertain, but this research demonstrates the clear benefits of physical activity to cognitive function in mammals, and it could have strong implications for the prevention and treatment of Alzheimer's disease and for age-related cognitive decline.

While Small and others investigate the relationships between glucose, exercise, neurogenesis, and memory at the molecular level, the epidemiological evidence in favor of physical exercise continues to roll in. A major longitudinal study released in June 2009 followed 2,500 people age 70 to 79 for a period of eight years, assessing their cognitive abilities several times over the course of the study period. "The majority of past research has focused on factors that put people at greater risk to lose their cognitive skills over time, but much less is known about what factors help people maintain their skills,"[21] says the study's author, Alexandra Fiocco of the University of California, San Francisco.

Many of the participants in the UCSF study showed an overall decline in cognitive function—setbacks ranging from mild to severe—but 30 percent of study participants showed no major changes, or else they improved their test performance over the years. When researchers analyzed the lifestyle choices that seemed to keep people sharp despite their age, regular physical exercise emerged as a major factor. People who exercised moderately to vigorously at least once a week were 30 percent more likely to maintain their cognitive skills than those who did not work out as often. Three other factors—education, not smoking, and staying socially active—also proved critical to cognitive health. Study participants with at least a high school education were nearly three times as likely to stay mentally sharp as those with less education, and older people with a ninth grade literacy level or higher were nearly five times as likely to maintain their mental skills as those with lower literacy levels. Nonsmokers were nearly twice as likely to stay sharp as those who smoked, and people who stayed socially active—people who worked or volunteered, or people who reported living with someone—were 24 percent more likely to maintain their cognitive skills into their later years.

"Some of these factors such as exercise and smoking are behaviors that people can change. Discovering factors associated with cognitive maintenance may be very useful in prevention strategies that guard against or slow the onset of dementia," Fiocco said. "These results will also help us understand the mechanisms that are involved in successful aging."[22]

Therapeutic forgetting

In March 2009, a team of Dutch researchers led by psychologist Merel Kindt made headlines when they successfully induced—then erased—the fear response in humans. Perhaps most astonishing, they managed to do it all in three short days.[23] The basic research question was this: What if, when a fearful memory is retrieved, we have a second chance to change the way it is stored in the brain, thereby easing its physical and emotional effects? It is a well-known neurological fact that it takes the brain time to store long-term memories. This delay is known as the labile phase, and during this phase, a memory is still unstable and susceptible to change. Conventional wisdom once held that if a memory was stored in long-term memory centers, the protein synthesis that recorded it was complete, permanent, and essentially bulletproof—except in cases where these memory centers were damaged through traumatic injury or conditions like Alzheimer's. But recent studies have shown that protein synthesis happens again during memory retrieval, raising the question of whether there is second opportunity to intervene chemically at this stage, thereby altering the restorage of the memory.

To test this idea, the Dutch researchers showed 60 undergraduate volunteers at the University of Amsterdam images of two different spiders. A mild electric shock accompanied one of the images but not the other. Coupling one of the spiders with a shock eventually produced a startle reflex in response to that image, whereas the students seeing the second image remained perfectly unruffled.

To measure the fear response, investigators placed electrodes beneath the subjects' eyes to record their blinking reflex (a reflex activated in the amygdala, the primal emotion center of the brain) and found that they had indeed successfully conditioned a fear response, just as researchers can teach lab rats to fear particular foods or places by shocking their little paws. This result was relatively unsurprising, but then on the second day of the experiment, some students received a dose of the beta-blocker propranolol—a drug that interferes with adrenaline absorption—whereas others received a placebo. Soon after administration of the drug, the students viewed the images again and the fear memory was reactivated. Earlier work had shown that especially vivid and durable memories form in the presence of adrenalin (specifically noradrenaline, or norepinephrine, the chemical form of adrenalin that acts on the brain), and that propranolol is one beta-blocker that acts on adrenal receptors in the brain as well as in the heart. "Propranolol sits on that nerve cell and blocks it," explains James McGaugh, psychologist at University of California Irvine and a leading investigator in the neurobiology of emotional memory. "So adrenaline can be present, but it can't do its job."[24] The Dutch researchers hoped that they would see a blunting of the fear response after reactivation of the memory in the presence of the beta-blocker.

The real test came on the third day, when students were shown the images again. This time, the physiological fear response to the first spider—the one associated with the electric shock—was eliminated in the group that received propranolol, but not in the placebo group, proving that administration of propranolol not only blunted, but erased, the fear response. Subjects remembered the image, and they remembered associating the painful stimulus with it, but that association no longer provoked a physiological or emotional reaction.

The investigators hope these results will contribute to a growing body of work on memory-based treatments for patients with anxiety disorders such as PTSD. Although it's unclear whether the findings can extend to severely traumatic memories or to memories that have been consolidated over a period of years or even decades, other research that is underway suggests that they might.

Roger Pitman, psychiatrist at Harvard Medical School, has been studying the therapeutic potential of propranolol for treating and preventing post-traumatic stress disorder since 2004, when he began to recruit patients arriving at the Massachusetts General Hospital emergency room after they had suffered serious traumas ranging from auto accidents and dangerous falls, to carjackings and sexual assaults.

Some patients were given propranolol; others were given a placebo. Kathleen Logue participated in the study after she was struck in the middle of a busy downtown Boston street by a bicyclist. "He just hit the whole left side of my body. And it seemed like forever that I was laying in the middle of State Street, downtown Boston,"[25] Logue remembered.

She volunteered for the study because she had developed PTSD eight years earlier as the victim of a carjacking and attempted rape. "I had a feeling that this one trauma, even though it was a smaller thing, would touch off all sorts of memories about that time I was carjacked," Ms. Logue said. "For eight months at least" after the assault, "every night before I went to bed, I'd think about it. I wouldn't be able to sleep, so I'd get up, make myself a cup of decaf tea, watch something silly on TV to get myself out of that mood. And every morning I'd wake up feeling like I had a gun against my head."[26]

Three months after being hit by the bike, Ms. Logue returned to the ER to recount her experience, and her story was compiled into a brief narrative and tape-recorded. A week later, her recorded story was played back to her. This protocol was followed for all 40 participants in the pilot study.

Upon hearing the recordings of their traumatic experiences, nearly half of patients who received the placebo showed physical symptoms of PTSD, whereas none of the patients who received the beta-blocker showed any such signs of physiological distress. These striking results led to a large-scale clinical trial with 128 patients funded by the National Institute of Mental Health, now in its final phase.[27]

Experiments like the one at Mass General provoke controversy because they tinker with the kinds of memories that many of us take to be at the core of our personal identity, and in 2003, the President's Council on Bioethics tackled the sticky subject in its report, "Beyond Therapy: Biotechnology and the Pursuit of Happiness." The council ultimately concluded that the practice of blunting or erasing human memories would be unethical, because the "use of such powers changes the character of human memory, by intervening directly in the way individuals 'encode,' and thus the way they understand, the happenings of their own lives and the realities of the world around them."[28]

The council's conclusions cut both ways. They wondered whether "dulling our memory of terrible things" would "make us too comfortable with the world, unmoved by suffering, wrongdoing, or cruelty?" They also wondered, however, whether blunting our recollections of "shameful, fearful, and hateful things" might diminish our ability to appreciate our happiest moments. "Can we become numb to life's sharpest sorrows without also becoming numb to its greatest joys?"[29]

Rebecca Dresser, a council member and professor of ethics in medicine at Washington University in St. Louis, expressed the majority sentiment when she said that the ability to suffer the "sting" of a painful memory is "where a lot of empathy comes from... There probably is some sting that we would rather not have as individuals, but it's good for the rest of us that others have it."[30]

David Magnus, director of Stanford University's Center for Biomedical Ethics, expressed concern that drugs like propranolol could be prescribed for trivial reasons. "From the point of view of a pharmaceutical industry, they're going to have every interest in having as many people as possible diagnosed with [PTSD] and have it used as broadly as possible. That's the reality of how drugs get introduced and utilized," Magnus says. "Our breakups, our relationships, as painful as they are, we learn from some of those painful experiences. They make us better people."[31]

Many brain scientists hold that emotional blunting, or even elimination, of traumatic memories would probably do much more psychological good than harm. "While memories are great teachers and obviously crucial for survival and adaptation," says Joe Tsien, whose research team has successfully erased memories in mice, "selectively removing incapacitating memories, such as traumatic war memories or an unwanted fear, could help many people live better lives."[32] Eric Kandel, professor of psychiatry and physiology at Columbia University, argues that although some difficult experiences are worth reliving, not all painful experiences are created equal. "Going through difficult experiences is what life is all about; it's not all honey and roses. But some experiences are different. When society asks a soldier to go through battle to protect our country, for instance, then society has a responsibility to help that soldier get through the aftermath of having seen the horrors of war."[33]

Roger Pitman finds arguments against propranolol for the treatment of PTSD to be incompatible with what we know are appropriate treatments for physical traumas, an inconsistency that he thinks reflects a deep-seated double standard. "Let's suppose that you have a person who comes in after a physical assault and they've had some bones broken, and they're in intense pain," Pitman says. "Should we deprive them of morphine because we might be taking away the full emotional experience? Who would ever argue that? Why should psychiatry be different? I think that somehow behind this argument lurks the notion that mental disorders are not the same as physical disorders. That treating them or not is more of an optional thing."[34]

Kathleen Logue, for her part, agrees that traumatized people should not be expected to live the rest of their lives with the debilitating effects of a traumatic memory as long as a safe and effective treatment is available. "It doesn't erase the fact that it happened," she says. "It doesn't erase your memory of it. It makes it easier to remember and function."[35]

This might be true for treatments that blunt the harmful physiological effects of traumatic memories, but what about deleting them altogether? New research on molecules responsible for storing memories suggests that this will be possible, perhaps sooner than we think. In October 2008, Joe Tsien with the Medical College of Georgia reported in the journal *Neuron* that his research team, in collaboration with scientists in Shanghai, had eliminated memories in mice by over-expressing a protein critical to brain function—known as αCaMKII—just as a memory was being recalled (during that second "labile" phase).[36] And two months later, researchers at SUNY Downstate Medical Center in Brooklyn shared findings that a molecule known to record memories, PKMzeta, preserves high-quality information about spatial and episodic memories but does not affect general abilities, suggesting that the substance could be used to erase particular memories—of trauma or addiction, for example—without harming our cognitive function.[37] "If further work confirms this view," said Andre Fenton, a professor of physiology and psychology and an author on the study, "we can expect to one day see therapies based on

PKMzeta memory erasure." Deleting negative memories, he says, "not only could help people forget painful experiences, but might be useful in treating depression, general anxiety, phobias, post-traumatic stress, and addictions."[38]

Research on PKMzeta and αCaMKII has been limited to animals so far, but the findings are likely to extend to humans, researchers say, and they hope that the new knowledge will lead not just to advances in the treatment of memory-related conditions such as PTSD and addictions, but to new treatments for dementia and age-related cognitive decline as well. "This is really the biggest target," said Todd Sacktor, head of the team at SUNY Downstate that discovered the importance of PKMzeta, "and we have some ideas of how you might try to do it, for instance to get cells to make more PKMzeta. But these are only ideas at this stage."[39]

If we begin to consider therapeutic memory erasure, what's to stop us from trying to insert new ones? What if we develop the know-how to manufacture and implant artificial memories? Perhaps we would limit the use of such technology to implanting happy memories that would support better physical and emotional health—in treating long-term depression or memory loss, for example. How would creating synthetic memories in these cases be less ethical than erasing actual ones?

If these questions make us feel a bit squirmy, perhaps it's because we are fast learning to fiddle with the most intimate and foundational parts of our historical selves, which might leave us wondering: What is left to call "me" after our most formative memories are peeled away? (There must be something left, right? It's whatever's feeling squirmy.)

In his book *Living in the Light of Death*, Buddhist meditation teacher Larry Rosenberg tells the story of a friend who was diagnosed with Alzheimer's disease. The man, who was also an experienced meditation practitioner, felt real sadness, fear, and embarrassment at the loss of his memory, but he learned how to apply his mental skills so that he could observe his lapses—many of them quite profound and frightening—without becoming overwhelmed by them. "He told me that when he awakens each morning," Rosenberg writes, "he has no idea of where he is or what he is supposed to do. He has learned to be mindful of such confusion and disorientation. He no longer panics. It passes, and he is able to get washed and ready for the day. So the episodes of memory loss became more and more manageable."[40]

What is this about us, exactly, that in the sudden and terrifying absence of our most intimate and formative memories chooses not to panic and decides to get up, brush our teeth, and be nice to people? Buddhist thinkers call this phenomenon "pure awareness," and they say that it is all that's left after we unpack the conditioned experiences we're taught throughout our lifetimes to believe in as "me." They also say it is the part of us that experiences true happiness, the part of us that really gets that the ultimate cause of happiness is kindness, and the only part of us that endures. We should take or leave these claims based on our own understanding and knowledge, of course, but the neural experience of self—and the emerging neural experience of no-self, or unselfconscious awareness—are two of the most intriguing and inspiring areas in neuroimaging research today.

7

Where am "I"? Experiences of self, other, and neither

We all have those "off" moments—sometimes entirely off days—when our inner experiences refuse to live up to our expectations, and we feel unwell, unsettled, or in pain. Times like these, we might wonder why "I just don't feel like myself." What does this mean, exactly? Who are these "myself" people—who do they *think* they are—telling us how we should or shouldn't feel?

Some days, of course, we're perfectly happy to identify with our sensations, emotions, and thoughts. Perhaps they feel pleasant, or they make us feel smart and principled, and we have no problem attributing them to "me." Other days we feel darker, more unproductive, and we might wonder what happened to that sharp, optimistic go-getter who inhabited our skull a mere 12 hours ago. Who do we become when we can't stand ourselves—when we've just done or said our worst, and the last thing on earth we want to do is own our petty thoughts and destructive urges? For that matter, who are we when we're our own biggest fan—when we're feeling brilliant and charmed and attractive, when apparently we can do or say no wrong?

If the real "self" would please step forward, we might ask it (for starters) where it lives, and if it wouldn't mind being at home when we need it.

There's a story about a time near the end of the Buddha's life when a follower asked him whether there is a soul separate from the body, and if so, what happens to it after death. The Buddha allegedly told his student that these were unknowable questions, and they would madden us and distract us from the vital spiritual work of

cultivating calm, wisdom, and balance in the face of life's joys and sorrows. He likened obsessing over such questions—instead of using
that time and energy to work with our own minds—to a man shot
with a poisoned arrow, refusing to allow the arrow to be removed
until he finds out everything he can about the weapon and the person
who shot him. "The man would die," the Buddha reportedly said,
"and those things would still remain unknown to him."[1]

It is also said, however, that the Buddha taught his students not to
take his (or anyone else's) word for anything, but instead to investigate the nature of experience using the powers of their own minds.
Apparently, two-and-a-half millennia later, we can't help ourselves:
Conscious awareness and the experience of "self" remain elusive
objects of scientific study, and they are among the hottest topics in
neuroimaging today. We are deeply curious about our own curiosity
and about every other mental phenomenon that contributes to a
sense of "me" and our place in the universe—imponderable or not.

"Here is this three-pound mass of jelly you can hold in the palm
of your hand," says neurologist V.S. Ramachandran, director of the
Center for Brain and Cognition at the University of California, San
Diego. "It can contemplate the vastness of interstellar space, it can
contemplate the meaning of infinity, and it can contemplate itself
contemplating the meaning of infinity. There is this peculiar recursive
quality that we call self-awareness which I think is the holy grail of
neuroscience, and hopefully someday we'll understand how that happens."[2]

Mysteries of the mind-brain connection notwithstanding, most
brain scientists subscribe to what Francis Crick has called the "astonishing hypothesis"—that all of our thoughts, sensations, emotions,
and beliefs—including our awareness of them—can be reduced to
electrical activity in the brain. Speaking for the majority opinion, Harvard psychologist Steven Pinker maintains that scientists have "exor-
cized the ghost from the machine not because they are mechanistic
killjoys but because they have amassed evidence that every aspect of
consciousness can be tied to the brain." Consciousness, he contends,
"turns out to consist of a maelstrom of events distributed across the
brain. These events compete for attention, and as one process out-
shouts the others, the brain rationalizes the outcome after the fact

and concocts the impression that a single self was in charge all along."[3]

Drawing another noise analogy, Michael Gazzaniga, director of the SAGE Center for the Study of Mind at the University of California, Santa Barbara, and former member of the President's Council on Bioethics, likens human consciousness to a pipe organ performing an elaborate and enthralling playlist. "What makes emergent human consciousness so vibrant," he says, "is that the human pipe organ has lots of tunes to play, whereas the rat's has few. And the more we know, the richer the concert."[4] He maintains that consciousness—the self that thinks it's the organist—is an "emergent property and not a process in and of itself."

Not so fast, say some people of faith who also happen to be scientists. They believe the apparent inseparability of subjective experience and neural activity says nothing about the existence (or not) of a soul—a musician playing the instrument of the brain. The subject of the soul, says Kenneth Miller, biologist with Brown University and a practicing Catholic, "is not physical and investigable in the world of science." When people ask him, "What do you say as a scientist about the soul?" he answers, "As a scientist, I have nothing to say about the soul. It's not a scientific idea."[5]

Buddhist thinkers, on the other hand, believe the idea of an atomistic, enduring, separate self is an illusion—one that causes a ton of unnecessary suffering—but they also believe that limiting our scope of inquiry to measurable physical phenomena is an incomplete strategy for understanding the nature of mind. This nature might or might not be reducible to physical phenomena, argues B. Alan Wallace, Buddhist philosopher and founder of the Santa Barbara Institute for Consciousness Studies, but if it is, we can't prove it yet. "We know so little about the mind and mind-brain interactions," he says, "it's way too premature to say that we know that the mind is nothing more than a function of the brain...the mind is nothing more than an emergent property of the brain. That might one day be demonstrated. Not yet. And therefore, if you want to understand the mind, let your primary mode of investigation be: *Observe it*."[6]

Wallace describes modern collaborations between Western neuroscience and ancient contemplative traditions as "bringing empiricism together with empiricism." Modern cognitive psychology, he argues, allows only for the few types of perception that are physical (such as sight, smell, taste, touch, sound). "They're all looking outwards into the world or into the body," he says. "But Buddhism has known for 2,500 years that we have another door of perception—of immediate access to certain aspects of reality—and that is mental perception."

Where Western science has come up short so far, he contends, is in elevating the objectively measurable, physical world to its primary object of study. "What makes the objective world more real than the subjective world?" he probes. "Just because you say so? Should we hold a church council and just decree that the objective things are somehow more real than the subjective? 'It's not real—it's only in your mind.'... What is more real than your sufferings and your joys, your hopes and fears, your thoughts, your aspirations, your memories, your sense of personal identity?"[7] Indeed, in light of the latest research on the direct impact of powerful mind states on our emotional lives and physical health, on our ability to make smart, compassionate choices and to build healthy relationships, it appears that nothing has more "real" implications for ourselves and our world than what we think and feel, whether or not we're aware of it, and how we act (or don't act) upon it.

Thanks to a meeting of empirical minds from two historically distinct traditions of inquiry—the modern scientific and the skilled contemplative—a peaceful revolution is unfolding inside the scanners. Experiments designed to watch the brain watch itself—imaging techniques combined with mindfulness techniques to study different kinds of awareness—are producing some of the most provocative images of the mind and its previously unappreciated powers. This practical alliance of curious people might never resolve "imponderable" questions about the existence of an enduring soul, but it is already changing our commonsense understanding of the "self" from a limited and static entity to a fluid and workable process, astonishingly receptive to the kinds of mental training that lead to health and happiness.

Decentralized self

One property of human consciousness emerging from the scanners—one fact that now seems inarguable—is that the neural activity associated with the subjective experience of "self" cannot be located in a single area of the brain. This finding matches what we now know about the entanglement of neural circuits associated with emotion, cognition, memory, and moral decision making. "The brain," says Richard Davidson, "does not respect the dichotomy of passions (emotion) and reason handed to us by the Greek tradition."[8] There is simply no isolated area of the brain we can point to as being purely emotional or purely cognitive. Likewise, there is no area of the brain we can point to as a single seat of consciousness. A more fitting analogy for the widespread neural "self" network might be a noisy family dinner table, and depending on who makes it home in time for the meal, who had a hard day or the best day ever, and who decides to shut up and pay attention, our subjective experience of self can vary wildly from moment to moment.

A recent study by researchers at the University of Toronto and Emory University used fMRI to contrast two distinct forms of self-reference: The "I" that is thinking, feeling, or doing in the present moment; and the explanatory "me," first conceptualized in Western psychology by the pioneering thinker William James.[9] The concept of "me," according to James, forms a coherent framework within which to make sense of the momentary "I" and its feelings, thoughts, and actions. "I," for example, am cooking dinner for my six-year-old son even though "I" am exhausted and would rather park it on the couch. Why don't "I" just toss him the car keys and tell him to go find his own food? That would never occur to "me" to do (most days), because parents always feed children and keep them safe, and I'm a parent. Kind people feed hungry people, and I'm a kind person. I've always fed my son and kept him safe; I'll do that until he's grown. This type of extended self-reference links past and present experiences with concepts and imagined future experiences, and it is dubbed "narrative focus" by the Toronto and Emory investigators. Recent imaging research has associated this type of thinking with activity in areas of the default network—the collection of circuits that come online when we're engaged in mind-wandering or ruminative thought and that hibernate when we are focused on more present-centered, goal-directed activity. In particular, the narrative "me" has been linked to activity in the medial prefrontal cortex (mPFC).

The other type of self-awareness—a momentary awareness of self that is centered in the present moment (termed "experiential focus" in the study)—has been more neglected in the neuroimaging litera-ture, and Norman Farb and Adam Anderson, in consultation with Zindel Segal and others, set out to explore the neural underpinnings of present-time awareness. "Most examinations of self-reference ignore mechanisms of momentary consciousness," they wrote, "which may represent core aspects of self-experience achieved earlier in development and may have evolved in earlier animal species." Awareness of the self in the moment might involve distinct neural processes from those involved in the narrative experience of self, but there's a hurdle to testing this theory. Left to its own devices, the typ-ical adult mind spends most of its time lost in mental time travel through the past and future—caught up in the extended "me" story—which appears as activity in the brain's default network, "potentially obscuring recruitment of distinct networks for more immediate self-reflection."[10]

The default state, notes Segal, has "a large emphasis on stimulus-independent thought—mind-wandering, chatter that cannot be cut off—and the deviation from that to a place where people aren't con-trolled by ongoing internal speech is a place where people are able to find some sort of calm and respite." Training in mindfulness medita-tion, Segal says, "is a way of helping people recognize that this chatter is omnipresent, but that they can actually learn ways of working with it so it doesn't end up ruling much of their emotional or even behav-ioral life."[11]

To tease apart the neural signatures of these two experiences of "self," the researchers imaged brains of two groups of people while they read eight lists of personality-trait adjectives, some positive (cheerful, mature, or talented, for example) and some negative (bit-ter, envious, or unkind). One group had registered for an eight-week mindfulness training class at St. Joseph's Hospital in Toronto but had not yet attended it; the other group had just completed the course. The investigators first trained both groups to employ narrative focus (with an emphasis on judging and evaluating what the personality-trait word means, and whether they thought it accurately described themselves) and experiential focus (with an emphasis on sensing thoughts and feelings and with no other goal than to notice moment-to-moment mental and physical experience). Inside the scanner, each trait word appeared on-screen for a period of several seconds, along with an icon to cue the participant to which type of attention to bring to the word.

For both groups, narrative self-focus produced heightened activity in cortical midline areas, including the mPFC (consistent with earlier studies of extended narrative self-reference), and in language areas heavily centered in the left hemisphere. Experiential self-focus, on the other hand, correlated with reduced activity along the cortical midline in both groups, and more dramatic and pervasive reductions appeared in the group that completed the mindfulness course. Participants trained in mindfulness also showed much greater activation in the right insula, an area that has been tightly linked to awareness of body sensations. The mPFC and right insula were strongly coupled in the novice group, a relationship that disappeared in the mindfulness group, suggesting that training in present-centered awareness might encourage a shift away from processing sensory information through the lens of the narrative self (as supported by activity in the mPFC).

"These results," the investigators wrote, "suggest a fundamental neural dissociation between two distinct forms of self-awareness that are habitually integrated but can be dissociated through attentional training: the self across time and in the present moment."[12] Indeed, the investigators noted, meditation practice has been associated with cortical growth in the right insula and other areas of the brain linked to sensory perception, suggesting that they might be strengthened through the regular exercise of moment-to-moment awareness, and "may represent the neural underpinnings of self-reference in the psychological present."[13] This apparent ability to disentangle our moment-to-moment conscious experience from an extended, narrative sense of self—and after a relatively brief training period of only eight weeks—should inspire those of us hoping to tip the scale toward happiness. Whereas ruminative, self-obsessed forms of thought that rehash the past or strategize about the future have been linked to mood and anxiety disorders and to increased morbidity and

mortality, more time spent being aware of the moment tends to correlate with feeling better, achieving more emotional balance, and
finding more enjoyment in our lives.[14]

Of two minds?

Perhaps the most convincing neurological evidence of the absence of a
single self running the show is the "split-brain" phenomenon, which
occurs in patients whose corpus callosum—the electrical connection
between hemispheres—has been totally severed. This injury to the
brain, as V.S. Ramachandran describes it, creates "two human beings
in one body, in one skull—two spheres of consciousness."[15]

Michael Gazzaniga is a foremost researcher on the bizarre effects
of this radical procedure, which is performed as a last resort on
patients with severe epilepsy. Gazzaniga has worked with patients like
Joe, who is aware of his surgery but says he can detect no difference
between his current and past states of consciousness. "My right hemisphere and my left hemisphere now are working independent of each
other, but you don't notice it," Joe reports. "You just kind of adapt to
it. It doesn't feel any different than it did before."[16] Despite Joe's
apparent ability to fuse his past and present mental lives into a contin-
uous experience, he is unable to consciously detect images to the left
of a dot in the center of a computer screen, as these images are
processed by his "disconnected" right brain rather than by his left
brain—the dominant hemisphere for language and speech. Though
he can't name the image or even consciously recognize that he's seen
it, he can close his eyes and draw it with his left hand, because that is
the hand controlled by the right hemisphere.

Patients like Joe, Gazzaniga believes, teach us that the mind is
made up of a constellation of "semi-independent agents, and that
these agents—these processes—can carry on a vast number of
activities outside of our conscious awareness." Throughout all of
this activity, there is some final system that constructs a coherent
theory to unify all these disparate elements. "That theory," says
Gazzaniga, "becomes our particular theory of our 'self' and of the
world."[17]

The concept of the brain's hypothesis-maker—dubbed the "interpreter" by Gazzaniga many years ago—grew out of his work with patients like Joe. Split-brain patients might seem "utterly and completely like you and me," Gazzaniga observes, "normal in every way, and yet we know that we can slip information into their silent right hemisphere and get them to do things."[18] The right hemisphere can be asked to get up and walk, for example, or to draw something, while the left brain is excluded from the conversation. "So there you are," Gazzaniga says. "Your hand is drawing something that your left brain doesn't know anything about, because it's your right brain that got the question, and the left hand draws it, but in fact the left brain is sitting there. It doesn't know anything about it."

After years of working with these patients, Gazzaniga and coworkers began to ask a simple question—"Why are you doing that?"—when split-brain patients performed an action in response to instructions given only to their right hemisphere. Why, for example, had their left hand just drawn that picture of a car? "Even though their hand had just drawn this thing," Gazzaniga says, "and the left brain really didn't know why, because we'd asked the disconnected right brain [to do it], they immediately came up with a story as to why they were doing a particular act. So they were interpreting behaviors that are coming out of them that were really generated by processes outside of the conscious awareness of the left hemisphere."

Gazzaniga believes that the hypothesis-maker resides in the left hemisphere and that it is intimately linked to our powers of language and expression. The interpreter, he says, is "the system that tells a story. What it's trying to do is seek the pattern that we're experiencing in life—whether it's emotional variations, whether it's actual behavior we produce—it's trying to build our personal narrative, our story line. And lots of stuff gets thrown into that." The interpreter is constantly generating explanations for our behavior—constantly coming up with stories about why we do things—even if there is no conscious motive for what we do at all because the activity is planned and executed outside of our conscious awareness. Whenever we are asked why we did something—even those of us whose brains and behavior could be considered within the normal, healthy range—"We cook up a story," Gazzaniga says, "and a rational view on why we're doing a particular thing."

The brain's story-maker turns out to have important implications for various kinds of neurological and psychiatric diseases where we can see the process working in quite exaggerated ways—in any disease, Gazzaniga says, "where you have a felt state that is exaggerated because of some kind of neurotransmitter state you're in." The mental state might feel painful or euphoric for the patient—"off" in some way—and they try to produce a story and a rationalization for why they might be feeling what they're feeling.

Gazzaniga gives the classic example of phobias, which can be generated out of a state of free-floating anxiety. A person has a sudden panic attack, and he tries to construct a theory about what caused it to happen. "Why am I having this awful feeling?" he might ask himself, and then he might misattribute the experience to his being in a particular place or setting. "'Well, it must be because I'm in this room, or I'm in this particular restaurant.' And you immediately build up a phobia for that sort of place," Gazzaniga says.

Hoping never to experience the unbearable feeling of free-floating anxiety again, the person concludes that it must have had something to do with a particular set of circumstances, circumstances that he goes out of his way to avoid in the future. But what generally happens with phobic patients, Gazzaniga explains, is that even if their anxiety can be tamed with medications, they will still suffer from the phobia itself, because it was generated by the interpreter as a logical explanation for a neurochemical experience. This is why most successful psychiatric interventions for phobias use a combination of treatments—drug therapy to correct the neurochemical imbalance responsible for the free-floating anxiety, and talk therapy to correct the false theory underlying the phobia.

If we don't buy Gazzaniga's theory that the "self" can be reduced to the interpreter—if we prefer to think of it as an eternal soul, or as a unique and separate mind somehow divisible from the brain—the existence of split-brain patients raises undeniably profound questions. To elucidate some of them, V.S. Ramachandran designed an experiment to test how the two hemispheres might produce different experiences of self. "Do they have different personalities?" he wondered. "What about their aesthetic preferences? Does one like blonds and the other like brunettes, for example? One like chocolate and the other like vanilla?"[19] Before asking questions of the different hemispheres in split-brain patients, he first had to train the right hemisphere to communicate. The right hemisphere does not possess the power of speech, but it can read and comprehend simple questions, and it can be trained to point to simple responses such as "Yes," "No," and "I don't know."

Ramachandran and his team then posed a series of questions to each hemisphere and compared the answers. "I asked, for example, 'Are you at Caltech?' and the right hemisphere pointed to 'Yes.' 'Are you on the moon?' It said, 'No.' 'Are you in California?' It said, 'Yes.' 'Are you asleep?' It said, 'No.' Then I said, 'Are you a woman?'" The patient was male, and he pointed to "Yes" and then started to laugh, proving, Ramachandran said, that the right hemisphere has a sense of humor. "Now comes the big question," he said. "What if you ask, 'Do you believe in God?'" When the question was asked of the right hemisphere, it answered "Yes." The left side, on the other hand, answered "No."

Here is a human being, notes Ramachandran, with one hemisphere that is an atheist and another that is a believer. This finding should have shaken the theological community to the core, he thinks, because it raises questions like, "If this person dies, what happens? Does one hemisphere go to heaven and the other go to hell? I don't know the answer to that."[20]

Ramachandran also highlights cases of anosognosia, a condition in which people who have suffered a disabling injury seem completely unaware of it. Stroke victims, for example, might suffer damage to the parietal lobe and paralysis on one side of the body, but not believe that they are paralyzed. "This person is quite intelligent, engages in a conversation about politics, can play chess with you and so on and so forth, but then you say, 'What about your arm? Does your left arm work?' and the patient says, 'Oh, it's fine! It works fine.' So here's a patient who is completely intelligent denying something perfectly obvious like the left arm being completely paralyzed." Ramachandran notes that in some extreme cases, patients will deny that the paralyzed limb belongs to them at all, saying, "Oh, it's my mother's hand," or "It's my father's hand."

In one such case, Ramachandran challenged the patient to touch her shoulder with the paralyzed arm (the arm she had claimed was not paralyzed). The patient responded by grabbing the arm with her able hand and lifting it to the shoulder. "Now *that's* amazing," Ramachandran says, "because it shows that *somebody* in there knows she's paralyzed."

No separate self

Studies of infant behavior in recent decades have confirmed what anyone watching a newborn clinging to a parent might guess—that none of us are born with an innate sense of ourselves as separate from our external world. The division between "self" and "other" is not hardwired in us from the start, but kicks in at some point in our development when a perceived separation between the physical body and the outside world becomes salient. With time and lots of practice, the concepts of "me" and "mine" start to stick, and they color our ever-widening circle of experience. *My* cookie, *my* side of the backseat in childhood becomes *my* intellectual property, *my* stock portfolio in adulthood.

In a letter to a friend who had lost a young son, Albert Einstein described the experience of self as somehow separate from the rest of reality as an "optical delusion" of consciousness, a delusion he likened to a prison. He said that to break free of this prison would require "widening our circle of compassion to embrace all living creatures and the whole of nature in its beauty." Acknowledging that this was a tall order, he also said that striving for this sense of connectedness is "in itself a part of the liberation and a foundation for inner security."[21] As usual, Einstein was on the mark: Study after study has linked feelings of connectedness and ties to a broader community with mental and physical health and longevity, whereas the emotional stress of loneliness and social isolation can be as harmful to our health as smoking or lack of physical exercise.[22]

What is it that makes the experience of "self" different from experiences in which self-consciousness is not present—when we are caring for others, for example, with no thought to our own needs or desires? Studies of memory have shown that we usually remember information related to our own lives better than other types of information, but psychologists have debated whether this happens because information about the self is processed in different parts of the brain, or simply because it is processed more frequently and solidified over time.

Recent brain imaging studies have indicated that at least one region of the brain, located in the medial prefrontal cortex (mPFC), lights up during self-referential processing and not during similar types of processing related to others (for example, when judging whether a particular adjective such as "trustworthy" describes oneself, versus whether it describes the president of the United States).[23] The mPFC also happens to be a hub of activity in the default network, and other imaging research has shown that default regions are only sparsely connected in young children as compared with adults,[24] suggestive of significant early-life plasticity in the networks supporting our narrative sense of self.

Unlike the architecture of the default network—and unlike our perceptual divisions between self and the external world—our ability to connect emotionally with experiences and feelings of others does appear to be ingrained from early childhood. Imaging studies of empathy show that the same brain regions that engage when a person feels pain also ignite when we imagine someone else experiencing the same type of pain, and a recent study by Jean Decety and his team at University of Chicago revealed this to be true even for children as young as seven. "Consistent with previous fMRI studies of pain empathy with adults," Decety and colleagues wrote, "the perception of other people in pain in children was associated with increased hemodymamic activity in the neural circuits involved in the processing of first-hand experience of pain."[25] Anyone who has heard a baby cry at the sound of another child in distress would probably suspect that infants—who, unfortunately, cannot be asked to hold still inside fMRI scanners—are hardwired for empathy from the get-go.

No self at all

Neurologist James Austin, author of *Zen and the Brain*, was catching the London Underground to a meditation retreat when a type of awareness overcame him that many experienced meditators describe as the insight of "no self." The habituated sense of "I," "me," and "mine" had evaporated, leaving in its wake a profound sense of inner peace and connectedness to the world. "Time was not present," he reported. "I had a sense of eternity. My old yearnings, loathings, fear of death and insinuations of selfhood vanished. I had been graced by a comprehension of the ultimate nature of things."[26]

Spiritual practitioners across a range of contemplative traditions describe similar experiences of God or insight—Navajo high priests see visions that dissolve the divide between the human world and the divine; Christian mystics experience unity with God through prayer; even hardnosed, skeptical empiricists feel moments of oneness or "flow" with the natural world. The source of these experiences remains elusive, though some researchers in the burgeoning field of neurotheology argue that they can be reduced to the effects of neurotransmitters or psychotropic chemicals in the brain. Others link experiences of heightened consciousness to structural or functional abnormalities, such as excessive activity in the temporal lobes.

"It has been known for a long time that some patients with seizures originating in the temporal lobes have intense religious auras—intense experiences of God visiting them,"[27] says V.S. Ramachandran, who has spent decades studying the phenomenon. He reports that these experiences are quite prevalent in patients with temporal lobe epilepsy, with 30 to 40 percent of patients experiencing periods of intense religious fervor. Even between seizures, patients will report spiritual experiences and feelings of direct connection with God. "Sometimes it's a personal God; sometimes it's a more diffuse feeling of being one with the cosmos," Ramachandran explains. "Everything seems suffused with meaning. The patient will say, 'Finally, I see what it is all really about, doctor. I really understand God. I understand my place in the universe—the cosmic scheme.'"

Ramachandran is fascinated by these episodes and why they are so common in patients with temporal lobe epilepsy, and he has designed experiments to test candidate theories. "One possibility," he says, is that "maybe God actually visits them." But if that theory is true, as a scientist he says he has no way of testing it. (He does express skepticism, however, that even a God who works in mysterious ways "would manifest herself in temporal lobe seizures, only in epileptics. But we don't know."[28])

Another possibility is that a seizure igniting the temporal lobe "somehow creates all kinds of odd, strange emotions in the person's mind," and the patient might interpret these experiences as visits from another world or from God. Perhaps, Ramachandran says, this is the only way the patient "can make sense of this welter of strange emotions going on in his brain."[29] This theory relies on the existence of

Gazzaniga's interpreter, or as Ramachandran describes it, "the left brain trying to spin a yarn to make things more consistent." If all this is true, then "God is the ultimate confabulation by the left hemisphere."

Still a third possibility is that the phenomenon has "something to do with the way in which the temporal lobes are wired up to deal with the world emotionally," Ramachandran says. As we interact with the outside world, we need some way of determining what's important to our survival and well-being, "what's emotionally salient, and what's relevant to you, versus something trivial and unimportant." It is vital to our survival as individuals, for example, that we be afraid of lions and tigers, and vital to our survival as a species that we be attracted to potential mates, so for most of us, images of predators and of scantily clad models tend to produce a measurable physiological response.

How does this process work in healthy brains? "We think what's critical is the connections between the sensory areas in the temporal lobes and the amygdala, which is the gateway to the emotional centers of the brain," Ramachandran says. "The strength of these connections is what determines how emotionally salient something is. And therefore you could speak of a sort of emotional salience landscape with hills and valleys corresponding to what's important and what's not important."

We all have slightly different "emotional salience" landscapes, he argues, and one theory about the religious experiences of patients with temporal lobe epilepsy is that repeated seizures might indiscriminately strengthen these pathways in the brain, and consequently produce the felt sense that every object in the world is imbued with deep significance. Finding emotional meaning in everything around us might generate the experience of being one with the entire cosmos or with God.

If this hypothesis were true, Ramachandran postulated, then images of trivial, everyday objects—not just lions and tigers and mothers—should produce a huge emotional jolt in patients with temporal lobe epilepsy, but this turned out not to be the case. Ramachandran employed a simple test of emotional and physiological arousal—the galvanic skin response—and found that patients showed the expected spikes in arousal when responding to images that should be frightening (violent scenes or images of predators, for example), and showed the normal lulls when responding to images of trivial objects. The patients' responses to religious iconography, however, told an entirely different story.[30] When they were shown a cross or the Star of David, or words like God or Jesus, there was a big jump in the galvanic skin response.

Ramachandran thinks these findings tell us that something special is going on in the temporal lobes—in temporal lobe epileptics, and perhaps in all of us to varying degrees—that makes humans in general more prone to religious belief. He suspects that this "something" could be a group of neurons firing in an abnormal manner, and that these neurons are simply hyperactive in people with temporal lobe epilepsy, "and hence the propensity for religious belief."

The scientific consensus about the origin of our basic spiritual tendencies is that they were probably selected for by regular old evolutionary forces. "Look at every society, every tribe in the world," Ramachandran notes. "They have some kind of religious belief." Perhaps similar experiences are seen across cultures because they support highly adaptive behaviors like cooperation in groups, mating rituals, and belief in hierarchy. "Religion," says anthropologist Scott Atran, "is a byproduct of many different evolutionary functions that organized our brains for day-to-day activity."[31] But religious believers, along with many neuroscientists, point out that evolutionary and neurobiological explanations for spiritual experiences can't contribute a thing to debates over the existence (or not) of God. As Ramachandran puts it, "None of this has any bearing on whether God really exists or not."

Neurologist Andrew Newberg of the University of Pennsylvania agrees. "The fact that spiritual experiences can be associated with distinct neural activity," he says, "does not necessarily mean that such experiences are mere neurological illusions."[32] Whatever the cause (or causes) of our spiritual encounters, centuries of anecdotal evidence support the notion that experiences of "higher consciousness" seem to occur when self-consciousness fades—when the concept of self as apart from the rest of reality stops making sense. Experienced meditators, for example, often describe conditions in which conscious awareness is present, but the sense of a separate self, or "self-consciousness," is nowhere to be found.

"With certain, particularly more advanced practitioners," says Richard Davidson, "the experience of no-self, along with very robust awareness, is something that does characterize certain kinds of meditation practice." There might be specific regions in the prefrontal and parietal cortices that are important for the representation of certain aspects of self, and "there may be decreased activation in those areas," he says, "while at the same time you get heightened activation in other prefrontal sectors that are critical for awareness."[33]

The experience of no-self is neurally distinct from the loss of an observing self in unconscious experience, which has been associated with widespread deactivation in prefrontal and parietal regions of the brain.[34] In sharp contrast to unconsciousness, the experience of no-self accompanied by strong awareness, particularly during open-presence meditation (a practice that involves watching the content of moment-to-moment experience in a nonreactive way), involves "heightened activation in most regions of the prefrontal cortex, not lower activation," says Davidson, but he notes that "there may be a selected decrease in activation in very specific subsectors that are important in aspects of self-representation." He emphasizes that this is extremely speculative territory, because little research to date has focused on the experience of no-self. "This is a very hot topic and one of great importance," he says. "We are trying to initiate a lot more research along these lines."[35]

Beyond self-concept

If we accept the neuroscientific (also Buddhist) premise that there is no executive thinker at the controls of the brain—no organist playing the organ—what is left, if anything, when the illusion of "me" falls apart?

Experienced meditation practitioners maintain that there is an enduring quality of mind beneath the illusion of a solid, separate self, and that this quality can be described as bare, nonjudgmental awareness. This awareness isn't "mine" or "yours"—it is simply awareness—which can observe the shifting nature of moment-to-moment experience with clarity, balance, and compassion. Matthieu Ricard, Tibetan Buddhist monk and scholar, likens basic consciousness to a mirror that allows all images to arise on it. "You can have ugly faces, beautiful faces," says Ricard. "The mirror allows that. But the mirror is not tainted, is not modified, is not altered by those images. Likewise, behind every single thought, there is bare consciousness—pure awareness—this is the nature." This basic nature cannot be distorted by passing emotions, Ricard says, because then evidence of them would always be present in the mind, "like a dye that would permeate the whole cloud." Experientially, we know this isn't the case, he points out, because we know we're not "always angry, always jealous, always generous."[36]

Implied in this mirror or "blank slate" analogy is the potential for anything to arise, and it frequently does. Meditators report being shocked at the range of thoughts and emotions that reveal themselves when they begin to calm their minds—mental and physical phenomena that had long been blocked from their conscious awareness by the illusion of "me." Awakening to the reality that anything can arise in our own minds—and by empathetic extension, everyone else's—is the foundation for compassion. "Awareness and compassion and kindness are very, very intimately related to each other," says Jon Kabat-Zinn. "In essence, they're not really different."[37]

Yongey Mingyur Rinpoche, lifetime practitioner of Tibetan meditation techniques and one of the 16 experts whose compassionate brain activity was studied at the University of Wisconsin, describes his practice in the fMRI scanner as one of nonreferential compassion, or "compassion beyond concept." To do this practice, first he allows his mind to rest on the "true nature" of all sentient beings, the basic goodness that "is the cause of loving-kindness." Suffering arises for us all, he says, when we are unable to recognize this true nature—this enduring potential beneath our impermanent, conditioned experiences that is "totally free from suffering, problems, obscurations."[38] Wishing for this painful mental block to dissolve is the essence of nonreferential compassion, and it was the meditation adepts' ability to rest their minds steadily on this wish that produced the stunning images of brain activity. Not only did Mingyur Rinpoche and the other experts register exceptionally powerful activity in regions linked to emotion sharing and empathy, they also showed heightened activation in areas involved in motor readiness. "They are poised to jump into action and do whatever they can to help relieve suffering,"[39] Davidson suggests.

All this well-wishing correlates with more conscious and benevolent behavior, and Davidson's team continues to probe how far the benefits of compassion extend beyond the individual practicing it. "It's our conjecture," Davidson says of these connoisseurs of kindness, "that their baseline state is not an ordinary baseline state." The Madison team has initiated an experiment with a highly trained Buddhist practitioner to test whether the mere presence of an extraordinarily compassionate being can alleviate the suffering of others.

"We know something about the circuitry that is involved in stress and in pain and in negative emotion," says Davidson, describing the experiment in which participants are exposed to a mildly painful stimulus or to pictures depicting others' suffering. "We do this under two conditions," he says. "One where they are in the presence of an experimenter, and one where they are in the presence of the monk."[40] He hypothesizes that being in the presence of someone "whose baseline state is one that exudes compassion" might reduce activation in circuits involved in negative emotions. If this true, it would match anecdotal reports of people feeling loved, safe, and inspired in the presence of a spiritual mentor, even in the absence of any direct interaction.

Paying attention

For those of us not destined for the monastic path, we might lead lives of relentless activity, lives that swirl around job and family obligations that barely leave us time to yawn, let alone notice what we're thinking or doing. The idea of ever securing enough quiet time or space to meditate might seem like a pipedream, and it might even feel selfish. Lucky for the frantic and guilt-prone among us, the practices of cultivating attention and positive emotion are extremely portable. We can take them with us to Little League practices and to the office, to the dinner table and to bed, into traffic jams and airport security lines and arguments. We aren't stealing time from the people we care about when we create time to be aware of their needs and consciously, carefully attend to them.

"There's nothing particularly Buddhist or mystical, Eastern or Western about paying attention," says Jon Kabat-Zinn. "It's something we're all capable of doing." Even the most stressed-out patients who walk into his clinic have the potential to attend to their bodies and minds in ways that can change their lives for the better. He makes a point of telling his patients, "From our perspective, there's more right with you than wrong with you, no matter what's wrong with you."[41]

Despite what the mail-order catalogues might have us believe, no new materialistic "stuff" is required to embark on this adventure— just our bodies and minds, our breath, and our basic wish for happiness. That basic wish is what leads us to focus our attention on *something* every moment of the day, but *what* we are attending to might or might not be good for us or for anyone else. "Even though the bewildered mind is untrained," says Sakyong Mipham Rinpoche, renowned teacher in the Shambhala Tibetan tradition, "it is already meditating, whether we know it or not. Meditation is the natural process of becoming familiar with an object by repeatedly placing our minds upon it."[42]

Sometimes that mental object is a kind wish for ourselves or for another person or animal, and sometimes it is whatever is unfolding in our physical environment or within the body. More often, though, our rushed and hectic lives set the stage for painful or destructive emotions to arise, without the companion knowledge that those emotions are already shifting, and might in fact be headed out the door. "When we get up in the morning and we're anxious about something," says Mipham Rinpoche, "anxiety becomes our view for the day: 'What about me? When will I get what I want?' The object of our meditation is 'me.'"[43] We identify with the anxiety, imbuing it with a sense of solidity and permanence, and we can easily get hooked by obsessive, ruminative thinking about what we believe will bring us happiness for the rest of the day, the rest of the week, the rest of our lives.

Luckily, there is always the opportunity to change the quality of our mental experience, says Matthieu Ricard, as long as we can remember that emotions are fleeting. "That is the ground for mind training," he says. "Mind training is built on the idea that two opposite mental factors cannot happen at the same time. You could go from love to hate, but you cannot at the same time, toward the same object—the same person—want to harm and want to do good. You cannot in the same gesture shake hands and give a blow."[44] If we make benevolent emotions our mental object, we'll have less time to rehearse destructive emotions and to strengthen the neural pathways associated with them. We'll step out of well-worn, fearful grooves and create new favorite paths through the mind, getting better at happiness and empathy, worse at anger and jealousy, and this shifting skill set will manifest itself in healthier brains, bodies, and relationships.

Research with highly trained meditators shows that there are clear benefits to longer stretches of practice, and if our work schedules and family duties allow for teacher-led retreats or even self-designed retreats, extended periods of practice may be in the cards. If retreats aren't practical, many local health centers now offer classes and services by clinicians trained in mindfulness-based stress reduction (MBSR) or mindfulness-based cognitive therapy (MBCT). But even if we find ways to swing extended periods of formal practice— periods lasting an hour or a weekend or longer—there will be plenty of noisy, crazy stretches in between when it will seem much more challenging to pay attention. Most meditation teachers suggest that brief, regular interludes of mindfulness practice are more fruitful than longer stretches every once in a while. Sakyong Mipham Rinpoche, for example, encourages people to start practicing in 10-minute increments, "without the goal of seeing immediate results," he says. "The sensible approach is to tell ourselves, 'I can still be irritated 90 percent of the time. But with 10 percent of my mind and heart, I'll try putting others first.'"[45]

Some days, it won't seem possible to carve out even a 10-minute space for practice, or we will simply be in no mood to try. When our minds and our lives seem to be racing each other, these are the times we most need to pay attention, and yet extended periods of quiet practice might not be within reach. On especially stressful days, the simple act of noticing a single breath can be a triumph; watching one angry thought come and go without acting on it might save a friendship. Moments like these add up to more aware, relaxed, compassionate lives, and three simple meditation practices—all of them tried and true for millennia, all of them now favorite subjects of brain imaging research—can create the conditions for mindful moments to arise at any time, in any situation, no matter how stressful.

Attention to the breath

Breath awareness is taught in a variety of meditation traditions to cultivate concentration, to calm the body, and to develop wisdom. "The focusing of attention on the breath is perhaps the most universal of the many hundreds of meditation subjects used worldwide," says Jack Kornfield, a leading American teacher of Buddhist mindfulness techniques and co-founder of the Insight Meditation Society in Barre, Massachusetts. Steadying the attention on the breath, he notes, is a central technique not only to yoga, Buddhism, Hinduism, Sufism, and other Eastern traditions, but to Christian and Jewish contemplative traditions as well. Breath awareness is a natural favorite of meditators from a wide range of spiritual traditions, he says, because it can "quiet the mind, open the body, and develop a great power of concentration. The breath is available to us at any time of day and in any circumstance. When we have learned to use it, the breath becomes a support for awareness throughout our life."[46] If we think of our habitually scattered mind as an untrained puppy, Kornfield says—good-natured but totally wild—we can think of breath awareness as one of the most powerful techniques at our fingertips for gently and kindly training the puppy to stay put and pay attention, rather than chasing after every single thought that arises.

The breath is taught as a natural source of key insights into the nature of experience—the related insights of impermanence and interconnectedness. The in-breath quickly turns to out-breath, and before we know it we're breathing in again; what we tend to think of as a solid, permanent physical process is actually a collection of sensations that are always shifting, always becoming something different. The second insight—the insight of interconnectedness—arises when we realize that no meaningful distinction can be made between the breath and the physical body, between the air coming in and out of the lungs and our body tissues, or between ourselves and the person sitting next to us who is sharing that air—even if that person just yelled at us after a hard day's work.

"We're all breathing," writes Larry Rosenberg in his well-known exposition of breath awareness teachings, *Breath by Breath*. "The instruction is just to know that we are, not in an intellectual sense, but to be aware of the simple sensation, the in-breath and the out-breath." This teaching goes against our lifetime of conditioning to want to control everything around us and inside us in the hopes of avoiding pain, staving off loss, and discovering the perfect recipe for happiness. "If we can learn to allow the breath to unfold naturally," he says, "without tampering with it, then in time we may be able to do that with other aspects of our experience: We might learn to let the feelings be, let the mind be."[47]

With practice, focusing the awareness on a single object like the breath—and gently bringing the attention back when it wanders—becomes more effortless, and the ability to attend to moment-to-moment experience becomes more stable. Focused awareness practices like breath meditation "create a sense of lightness and vigor" in highly trained practitioners, Davidson, Lutz, and colleagues note, and "a significant decrease in emotional reactivity." Advanced concentration skills also correlate with a need for less sleep[48] and improved ability to attend to objects that are embedded in a rapid stream of stimuli,[49] suggesting that focused awareness practice leads to a more efficient distribution of limited brain resources and to reduced mental fatigue.

Open awareness

After we settle the mind with attention to the breath (or some other focused awareness practice), another set of practices available to us involves the expansion of our field of awareness to include all experience. These practices, also called objectless attention or open monitoring practices, involve watching the content of moment-to-moment experience in a nonreactive way—not judging the content as good or bad, desirable or undesirable, but rather noticing it as it unfolds within and around us.

"Thoughts, feelings, and sensations may come and go, but you just observe them," says Yongey Mingyur Rinpoche. "You pay light and gentle—or what we in the Buddhist tradition refer to as 'bare'—awareness to them, as you rest with an 'Ahh,' simply open to the present moment." He likens the "Ahh" feeling to the relief we experience after finishing a tiring or difficult task. "Just let go and relax," he says. "You don't have to block any thoughts, emotions, or sensations that arise, but neither do you have to chase them."[50] According to Davidson, Lutz, and colleagues, meditators who practice this relaxed, open form of awareness report "more acute, but less emotionally reactive, awareness of the autobiographical sense of identity that projects back into the past and forward into the future."[51] Open awareness practices tend to produce a heightened sensitivity to the body and the physical environment, accompanied by a decrease in the emotional reactivity that can worsen mental distress.

In their MBCT course for depression, Williams, Teasdale, Segal, and Kabat-Zinn introduce a practice they call the "three-minute breathing space," an exercise that alternates between focused and open awareness to relax the body, to become more aware of feelings and thoughts, to gather concentration through breath awareness, and to apply this enhanced concentration and relaxation to a more open field of awareness. The three-minute breathing space, its creators say, "can be deployed anytime, anywhere, for the duration of one or two breaths, to 5–10 minutes, as conditions permit."[52] This practice turns out to be a favorite for many MBCT course participants, and preliminary results suggest that PTSD sufferers find it a useful tool as well. The practice involves three basic steps: first, "becoming aware" of thoughts and feelings as they arise in the body and the mind, and acknowledging their presence; second, "gathering" our attention by focusing on the sensations of the breath and by feeling the expansion and contraction of the body, using the breath to anchor us in the present; and third, "expanding" our field of awareness to include the whole body and our facial expression. If we become aware of any feeling of tension or discomfort, we can breathe into it and meet it with acceptance. The MBCT authors suggest that we might want to reassure ourselves on the outbreath, "'It's okay...whatever it is, it's already here: Let me feel it.'"[53]

This handy technique combines many of the strengths of focused and open-field awareness in a concise and portable practice, and can prove especially useful in situations where we need to decide quickly how to handle uncomfortable or painful feelings, emotions, and thoughts. "In such situations, when low mood threatens to overwhelm us, the breathing space allows us to steady ourselves. It allows us to see clearly what is happening through direct, experiential knowing. It provides a place from which we can choose mindfully what next steps are required by the situation we find ourselves in."[54]

Compassion and loving-kindness

These twin objects of meditation are the wish for the end of suffering (compassion) and the wish for enduring happiness (loving-kindness— or *metta* in Pali, *maitri* in Sanskrit). These practices take many forms, including the recitation of phrases ("May I be free from suffering and the causes of suffering," for example, or "May my dear friend experience clarity of mind and kindness of heart," or "May my children be safe and joyful," and so on), eventually extending our well-wishing to emotionally neutral relationships, difficult relationships, and finally to all beings. The practices can also take nonverbal forms, such as using mental images of suffering to generate compassion, or mental images of "difficult" people as children to generate kindness. Yet another form, nonreferential compassion or loving-kindness (as practiced by Yongey Mingyur Rinpoche and the other monks in the fMRI scanner), can be described as empathy and well-wishing directed toward all sentient beings, in the absence of any explicit object of concentration.

Unlike conditional forms of love for oneself and others in which we expect to get something out of the bargain, loving-kindness "doesn't mean getting rid of anything," writes the well-known American teacher in the Tibetan tradition, Pema Chödrön. "*Maitri* means that we can still be crazy after all these years. We can still be angry after all these years. We can still be timid or jealous or full of feelings of unworthiness. The point is not to try to change ourselves. Meditation practice isn't about trying to throw ourselves away and become something better. It's about befriending who we are already."[55]

Loving-kindness is often described as an unconditional, open, and unobstructed form of love—caring for things as they are right now, rather than restricting our ability to care for ourselves and others with our ideas about how things should be. Sharon Salzberg, a leading American teacher of the practice and co-founder of the Insight

Meditation Society in Barre, Massachusetts, points out that the literal meaning of the word *metta* is friendship. "It means developing the art of friendship both toward ourselves—and not just those parts of ourselves we like and we proudly present to the world—but all aspects of ourselves," she says, "and then friendship toward all of life. What it really means is an acknowledgement of connection."[56] Approaching the practice as if we're cultivating friendships, we capture the spirit of acceptance that is a necessary component of the emotion. "A friend may disappoint us; she may not meet our expectations," Salzberg says, "but we do not stop being a friend to her. We may in fact disappoint ourselves, may not meet our own expectations, but we do not cease to be a friend to ourselves."[57]

Traditional Buddhist texts distinguish loving-kindness from other emotions that are easily confused with it, like passion, obsession, and sentimentality. Unlike these experiences we often think of as love, practicing loving-kindness means "we open continuously to the truth of our actual experience, changing our relationship to life," Salzberg says. "Metta—the sense of love that is not bound to desire, that does not have to pretend that things are other than the way they are—overcomes the illusion of separateness, of not being part of a whole."[58]

We all have moments in our lives, she says, when we see this connectedness clearly, when we're hit by a wave of insight—maybe it feels like a fine-tuned sensitivity or an engaged conscience—and we understand that we are all deeply vulnerable and that we are all at our most basic level just seeking happiness. In these moments of heightened clarity, Salzberg says, "we understand happiness really differently, we see our fears for what they are, we realize we might die in five minutes—*something* happens, and we realize we're not all that different."[59] It is often a personal tragedy—or the threat of one, she says—that brings realization of this connectedness. One moment we're going along, business as usual, seeing all of our experiences through the lenses of self and other—us and them—and then something happens and "all of those assumptions get challenged in one moment.... You get one phone call, and it's a different life. And that's true for *everybody*. It's not just true for some."

Many meditators report that the practices of compassion and kindness resonate more deeply than other concentration practices, because the object of mindfulness is indistinguishable from the quality of mind they wish to cultivate. Others can't stand the practice at first, reporting that it feels insincere or fake, or perhaps Pollyannaish in the face of extreme suffering in the world. Our personal histories and passing moods will surely influence which mental practices work best for us at any given moment. There will be days when consciously cultivating kindness won't feel like our "thing," perhaps because it feels too forced or too emotionally taxing. It is said that this is why the Buddha taught not one, not three, but hundreds of practices for training the mind to be clearer, kinder, more relaxed, and more attentive to the moment—the only time when the conditions for happiness actually come together.

All we can ask of ourselves is to try a technique or two and see what works, and then see what works tomorrow or a year from now or decades from now—when we are enthusiastic, when we lack a sense of purpose, when we feel unstoppable, when we experience a tragedy, when we are dying. For in-depth instructions on these three (and a wealth of other) practices, the "Resources" section at the end of this book provides information on retreat centers, MBSR and MBCT programs, and free teacher talks available on the Web, as well as a list of popular books on self-retreats and on daily sitting, walking, eating, and mindful living practices.

Think "human"

Why bother with such heady, impractical questions as "What is the 'self?'" Whatever the answer, our self (or our not-self, as the case may be) is going to need to roll out of bed in the morning, pour itself a bowl of cereal, and attend to its responsibilities. Most days, our "self" does just fine not overthinking things. It powers through the stress of the day, taking care of itself and the people it loves, perhaps taking care of a few people it doesn't even know. It saves for its retirement and its children's futures. If it's lucky, it gets to do a job that it loves and it learns new things along the way. What more could it ask from life?

Each fundamental question of interest in science, says Jon Kabat-Zinn, "really comes down to 'Who are we?' and 'What is our place in the universe?'"[60] Many brain scientists already have their pet answers, some suggesting that what makes humans unique among beasts is our capacity for awareness; others pointing to our ability to regulate powerful emotions; still others emphasizing our ability to create and appreciate meaning and art. These so-called "human" qualities are all obviously and intimately linked to the brain; no one places particular importance on our hairless tummies or lack of a wagging tail. As interesting as what makes us different from other sentient beings, however, is what makes us precisely the same. "Despite what the French intellectuals say," notes Matthieu Ricard, "it seems that no one wakes up in the morning thinking, 'May I suffer the whole day,' which means that somehow, consciously or not, directly or indirectly, in the short or the long term, whatever we do, whatever we hope, whatever we dream somehow is related to a deep, profound desire for well-being or happiness."[61]

In light of this inborn drive for happiness, it is understandable that we would utilize all the human faculties at our disposal—our ability to cultivate awareness, to regulate our emotions, to appreciate love and beauty, to care for others, to operate fMRI machines—in the active pursuit of well-being. Contemplative practices allow us to show up for our lives, teaching us how to pay attention to our experiences while we can actually experience them. Brain imaging research gives us the complementary ability to watch that inner life unfold in real time on-screen—a power that affords us unprecedented glimpses into the flexibility and trainability of our minds.

Yongey Mingyur Rinpoche sees the great potential in brain imaging research to help us recognize who we are without "always looking at material phenomena." Watching pictures of the brain respond to experience, he says, might help us to recognize, "'Oh! Maybe the source of happiness—the cause of happiness—not only exists in outer material, but exists in the mind.'" Wisdom does not always face outward, he says, but also faces "inward to see your true nature, the positive quality of your mind," and to see how this quality translates into healthy actions and relationships.[62]

Our internal experiences, along with our relationships and our actions in the world, will add up to what we look back on and call our life. A profoundly liberating alliance between scientific and contemplative traditions is opening up a wealth of future possibilities for us to change our lives and others' lives for the better, by learning to attend to our internal experiences and our reactions to them, moment by moment, with kindness.

Our curious identity

Questions such as, "Who are we?" "What does it mean to be alive?" "What does it mean to be human?" and "What will it take to be happy?" have fed the human imagination for as long as we've had one. Curious people through the ages have gnawed these questions from every available angle—sometimes turning inward to mental activity for answers, sometimes turning outward to observe fellow humans interacting with the world. Our newly discovered power to picture the brain in real time gives us a fresh angle that lets us watch our inner subjective experience unfold in unison with objectively measurable brain activity. "All of these questions that philosophers have been studying for millennia," says V.S. Ramachandran, "we scientists can begin to explore by doing brain imaging and by studying patients and asking the right questions."[63]

A patient shows no external signs of awareness. Researchers ask her to imagine playing tennis or walking through the familiar rooms of her home while they watch her brain engaged in each task. They discover that she is, in fact, aware of herself and her surroundings, able to follow directions with a demonstrable sense of purpose. Handing an otherwise powerless person a communication device in the form of functional MRI—asking her to actively participate in the experiment with all her reserve abilities—these researchers are able to uncover an aware, productive human life that otherwise might have gone unnoticed. In so doing, they expose the potential for brain imaging technology to extend the powers of speech and autonomous decision making to patients living in an otherwise locked-in state of consciousness, and they foreshadow the potential for imaging technology to define the meaningful limits of awareness and therefore of sentient human life—perhaps one day allowing us to know when to hold onto the life of a loved one and when to let it go.

Across the ocean, a team of investigators asks people with localized brain damage a series of moral questions, discovering that a particular region of the brain is key to "normal" moral reasoning. A second team reveals that damage to a different area of the brain wipes out the uniquely human ability to appreciate abstraction and creative metaphor. And a third team finds that dysfunction in the circuitry of pain and reward interrupts our instinctive human ability to empathize with others.

Our curiosity doesn't stop with figuring out what is broken; we want to fix it. Another group of investigators uses imaging technology to teach chronic pain patients how to find relief by regulating their brain activity using realtime fMRI feedback, foreshadowing the potential use of neuroimaging therapy to heal other debilitating conditions affecting well-delineated brain systems.

At the other end of the wellness spectrum, teams of researchers demonstrate the enhanced physical health, emotional balance, and cognitive acuity of graduates of mindfulness-based stress reduction classes and—at the most contented extreme—meditation experts who spend their lives cultivating positive qualities of mind. By imaging the brain while we ask it smart questions, and by linking the answers to observable neural activity, scientists are able to identify what goes wrong when critical aspects of normal human functioning are lost, and what goes precisely right when we train our minds in ways that maximize our potential for health and happiness. These experiments enhance our knowledge of what it means to be human and what it means to be well, laying the groundwork for targeted treatment of mental injuries and illnesses, and for improved quality of life for the fittest among us who want to do better than simply power through the stress. We'd like to thrive instead of merely survive, and we'd like others to experience true wellness, too.

Some tendencies of mind are conducive to flourishing—these are the qualities we'd like to reinforce or strengthen—and others haven't led us toward health and happiness. Whatever our self-maintenance "to do" lists look like, cutting-edge brain science is uncovering the tools we need to do the work. Among the most efficient are regular physical exercise and relaxation, systematic cultivation of positive emotions, engagement with strong social support networks, frequent cognitive challenges, and enhanced awareness of our internal states from moment to moment.

Everything we learn about the resilient, flexible brain challenges the nature of our common-sense understanding of ourselves as atomistic, limited identities bouncing off one another, and reveals that each time we interact, we change each other's brains, and that each time we respond to a thought or emotion, we change our own. A new science of mind—a hybrid of external and internal study—lets us watch our experiences unfold in real time while linking them directly to the experiences of others in objectively measurable ways. This profound new technical power affords us precious insight into the fluidity of our mental lives and into the productive ways in which we can direct and harness that flow.

Our old friend curiosity has crafted a fresh set of tools with which to watch itself, and they are showing us in stunning detail and color that the human potential for understanding, happiness, and mental freedom is astonishingly vast—perhaps limitless. Whatever age we happen to be, whatever our accumulated life experience, each of us wields the power to consciously improve the quality of our mental lives, and by extension, our physical health and our interactions with others. The more we observe the landscape of the mind, the clearer we see that whether the "other" is a loved one or a perfect stranger, a PTSD sufferer or a Buddhist monk, a healthy 7-year-old or a 70-year-old chronic pain patient, our shared mental potential is at least as defining as any features that set us apart.

Resources

Brain imaging and health on the Web

For the latest information on the research and treatment topics highlighted in this book, the following Web sites provide free, searchable text and links to other relevant sources. Many also include interactive features and audiovisual content, such as video clips and podcasts.

Alzheimer's Foundation of America, www.alzfdn.org/

American Pain Society, www.ampainsoc.org/

American Psychological Association, www.apa.org/

Center for Mindfulness in Medicine, Health Care, and Society, University of Massachusetts Medical School, www.umassmed.edu/cfm/index.aspx

Mayo Clinic, www.mayoclinic.com/

Mind and Life Institute, www.mindandlife.org/

Mindfulness-Based Cognitive Therapy (MBCT), www.mbct.com/

Mindfulness-Based Stress Reduction (MBSR), www.umassmed.edu/cfm/mbsr/

National Institutes of Health (NIH), www.nih.gov/. Links to individual institutes are available at www.nih.gov/icd/.

The following institutes fund much of the research featured in this book:

- National Institute of Mental Health (NIMH),
  www.nimh.nih.gov/
- National Institute on Drug Abuse (NIDA),
  www.nida.nih.gov/
- National Institute on Neurological Disorders and Stroke
  (NINDS), www.ninds.nih.gov/
- National Institute on Aging (NIA), www.nia.nih.gov/

Public Library of Science (PLoS), www.plos.org/

Science Daily, www.sciencedaily.com/

Society for Neuroscience, www.sfn.org/

Teachings on mindfulness-based therapies and practical meditation techniques

Charlotte Joko Beck, *Everyday Zen: Love and Work* (New York: HarperOne, 2007).

Sylvia Boorstein, *Don't Just Do Something, Sit There: A Mindfulness Retreat with Sylvia Boorstein* (New York: HarperOne, 1996).

Pema Chödrön, *When Things Fall Apart: Heart Advice for Difficult Times* (Boston: Shambhala, 2000).

Joseph Goldstein, *Insight Meditation: The Practice of Freedom* (Boston: Shambhala, 2003).

Jon Kabat-Zinn, *Full Catastrophe Living: How to Cope with Stress, Pain, and Illness Using Mindfulness Meditation* (London: Piatkus Books, 2001).

Jack Kornfield, *A Path with Heart: A Guide Through the Perils and Promises of Spiritual Life* (New York: Bantam, 1993).

Larry Rosenberg, *Breath by Breath: The Liberating Practice of Insight Meditation* (Boston: Shambhala, 2004).

Sakyong Mipham, *Turning the Mind into an Ally* (New York: Riverhead Books, 2003).

Sharon Salzberg, *Lovingkindness: The Revolutionary Art of Happiness* (Boston: Shambhala, 2008).

Thich Nhat Hanh, *Peace Is Every Step: The Path of Mindfulness in Everyday Life* (New York: Bantam, 1992).

Mark Williams, John Teasdale, Zindel Segal, and John Kabat-Zinn, *The Mindful Way Through Depression: Freeing Yourself from Chronic Unhappiness* (New York: Guilford Press, 2007).

Yongey Mingyur Rinpoche and Eric Swanson, *The Joy of Living: Unlocking the Secret and Science of Happiness* (New York: Harmony Books, 2007).

Talks by meditation teachers are available for free download from the following Web sites:

- Dharma Seed, www.dharmaseed.org/
- Plum Village Practice Center, www.plumvillage.org/
- San Francisco Zen Center, www.sfzc.org/
- Shambhala International, www.shambhala.org/

Notes

Introduction

[1]Aristotle, *On the Parts of Animals, Book II*, translated by William Ogle (Adelaide, South Australia: University of Adelaide Library eBooks@Adelaide, 2007).

[2]Aristotle, *On the Generation of Animals, Book II*, translated by Arthur Platt (Adelaide, South Australia: University of Adelaide Library eBooks@Adelaide, 2007).

Chapter 1

[1]Adrian M. Owen, Martin R. Coleman, Melanie Boly, Matthew H. Davis, Steven Laureys, and John D. Pickard, "Detecting Awareness in the Vegetative State," *Science* 313 (8 September 2006): 1402.

[2]Benedict Carey, "Mental Activity Seen in a Brain Gravely Injured," *New York Times*, 8 September 2006.

[3]Adrian M. Owen, Martin R. Coleman, Melanie Boly, Matthew H. Davis, Steven Laureys, Dietsje Jolles, and John D. Pickard, "Response to Comments on 'Detecting Awareness in the Vegetative State,'" *Science* 315 (2 March 2007): 1221c.

[4]Daniel L. Greenberg, "Comment on 'Detecting Awareness in the Vegetative State,'" *Science* 315 (2 March 2007): 1221b.

[5]William Saletan, "The Unspeakable: Buried Alive in Your Own Skull," *Slate*, 12 September 2006.

[6]Benedict Carey, "Mental Activity Seen in a Brain Gravely Injured," *New York Times*, 8 September 2006.

[7]Steven Laureys, "Eyes Open, Brain Shut," *Scientific American*, May 2007, 37.

[8]Ibid., 32.

[9]Benedict Carey, "Inside the Injured Brain, Many Kinds of Awareness," *New York Times*, 5 April 2005.

[10]Benedict Carey, "Mental Activity Seen in a Brain Gravely Injured," *New York Times*, 8 September 2006.

[11]Rebecca Morelle, "I Felt Trapped Inside my Body," BBC News, 7 September 2006.

[12]Robert Langreth, "Twilight Zone: Can Medicine Help Revive Brain-Damaged Patients Stuck in the Netherworld Between Coma and Consciousness?" Forbes.com, 4 October 2004, www.forbes.com/business/global/2004/1004/060.html.

[13]Rebecca Morelle, "I Felt Trapped Inside my Body," BBC News, 7 September 2006.

[14]Ibid.

[15]Ibid.

[16]Benedict Carey, "Mental Activity Seen in a Brain Gravely Injured," *New York Times*, 8 September 2006.

[17]Eelco F.M. Wijdicks, "Minimally Conscious State vs. Persistent Vegetative State: The Case of Terry (Wallis) vs. the Case of Terri (Schiavo)," *Mayo Clinic Proceedings* 81, no.9 (September 2006): 1157.

[18]Benedict Carey and John Schwartz, "Schiavo's Condition Holds Little Chance of Recovery," *New York Times*, 26 March 2005.

[19]Abby Goodnough, "Schiavo Autopsy Says Brain, Withered, Was Untreatable," *New York Times*, 16 June 2005.

[20]Robert Langreth, "Twilight Zone: Can Medicine Help Revive Brain-Damaged Patients Stuck in the Netherworld Between Coma and Consciousness?" Forbes.com, 4 October 2004.

[21]Kirk Payne, Robert M. Taylor, Carol Stocking, and Greg A. Sachs, "Physicians' Attitudes About the Care of Patients in the Persistent Vegetative State: A National Survey," *Annals of Internal Medicine* 125, no. 2 (July 1996): 104–110.

[22]Eelco F.M. Wijdicks, "Minimally Conscious State vs. Persistent Vegetative State: The Case of Terry (Wallis) vs. the Case of Terri (Schiavo)," *Mayo Clinic Proceedings* 81, no.9 (September 2006): 1155.

[23]Ibid.

[24]Benedict Carey, "Inside the Injured Brain, Many Kinds of Awareness," *New York Times*, 5 April 2005.

[25]N.D. Schiff, D. Rodriguez-Moreno, A. Kamal, K.H. Kim, J.T. Giacino, F. Plum, J. Hirsch, "fMRI Reveals Large-Scale Network Activation in Minimally Conscious Patients," *Neurology* 64 (February 2005): 514–23.

[26]Robert Langreth, "Twilight Zone: Can Medicine Help Revive Brain-Damaged Patients Stuck in the Netherworld Between Coma and Consciousness?" Forbes.com, 4 October 2004.

[27]Nick Chisholm and Grant Gillett, "The Patient's Journey: Living with Locked-in Syndrome," *BMJ* 331, no. 7508 (9 July 2005): 94.

[28]TV New Zealand, "Struggle Street," *20/20*, 20 June 2007, www.youtube.com/watch?v=7tG4BCAT9iA.

[29]Chisholm and Gillett, "The Patient's Journey: Living with Locked-in Syndrome," *BMJ* 331, no. 7508 (9 July 2005): 95.

[30]José León-Carrión, Philippe van Eeckhout, María del Rosario Domínguez-Morales, Francisco Javier Pérez-Santamaría, "The Locked-in Syndrome: A Syndrome Looking for a Therapy," *Brain Injury* 16, no. 7 (2002): 571-582.

[31]Steven Laureys, Frédéric Pellas, Philippe Van Eeckhout, Sofiane Ghorbel, Caroline Schnakers, Fabien Perrin, Jacques Berré, Marie-Elisabeth Faymonville, Karl-Heinz Pantke, Francois Damas, Maurice Lamy, Gustave Moonen, Serge Goldman, "The Locked-in Syndrome: What Is It Like to Be Conscious but Paralyzed and Voiceless?" *Progress in Brain Research* 150 (2005): 505.

[32]Chisholm and Gillett, "The Patient's Journey: Living with Locked-in Syndrome," *BMJ* 331, no. 7508 (9 July 2005): 95.

[33]Emanuela Casanova, Rosa E. Lazzari, Sergio Lotta, Anna Mazzucchi, "Locked-in Syndrome: Improvement in the Prognosis After an Early Intensive Multidisciplinary Rehabilitation," *Archives of Physical Medicine and Rehabilitation* 84, no. 6 (June 2003): 862-867.

[34]Steven Laureys et al., "The Locked-in Syndrome: What Is It Like to Be Conscious but Paralyzed and Voiceless?" *Progress in Brain Research* 150 (2005): 505.

[35]TV New Zealand, "Struggle Street," *20/20*, 20 June 2007, www.youtube.com/watch?v=7tG4BCAT9iA.

[36]Chisholm and Gillett, "The Patient's Journey: Living with Locked-in Syndrome," *BMJ* 331, no. 7508 (9 July 2005): 97.

[37]TV New Zealand, "Struggle Street," *20/20*, 20 June 2007, www.youtube.com/watch?v=7tG4BCAT9iA.

[38]Chisholm and Gillett, "The Patient's Journey: Living with Locked-in Syndrome," *BMJ* 331, no. 7508 (9 July 2005): 97.

[39]Steven Laureys et al., "The Locked-in Syndrome: What Is It Like to Be Conscious but Paralyzed and Voiceless?" *Progress in Brain Research* 150 (2005): 495.

[40]TV New Zealand, "Struggle Street," *20/20*, 20 June 2007, www.youtube.com/watch?v=7tG4BCAT9iA.

[41]Chisholm and Gillett, "The Patient's Journey: Living with Locked-in Syndrome," *BMJ* 331, no. 7508 (9 July 2005): 97.

[42]M. Bruno, J.L. Bernheim, C. Schnakers, S. Laureys, "Locked-in: Don't Judge a Book by Its Cover," *Journal of Neurology, Neurosurgery, and Psychiatry* 79, no. 1 (January 2008): 2.

[43]"Vegetative Patient 'Communicates,'" BBC News, 7 September 2008.

[44]Adrian Owen et al., "Detecting Awareness in the Vegetative State," *Science* 313 (8 September 2006): 1402.

[45]Alain L. Sanders, Jerome Cramer, and Elizabeth Taylor, "Whose Right to Die?" *Time*, 11 December 1989.

Chapter 2

[1]Richard Davidson, interviewed by Miriam Boleyn-Fitzgerald, 26 June 2008, transcript of audio recording.

[2]Richard Davidson, "Shaping Your Child's Brain," talk given at Appleton East High School (Appleton, Wisconsin), 13 May 2008.

[3]Katherine Ellison, "Mastering Your Own Mind," *Psychology Today* (1 September 2006).

[4]Ibid.

[5]Elaine D. Eaker, Lisa M. Sullivan, Margaret Kelly-Hayes, Ralph B. D'Agostino, and Emelia J. Benjamin, "Anger and Hostility Predict the Development of Atrial Fibrillation in Men in the Framingham Offspring Study," *Circulation* 109 (2004): 1267–1271.

[6]Childhelp, "National Child Abuse Statistics," 2006, www.childhelp.org/resources/learning-center/statistics.

[7]National Coalition Against Domestic Violence (NCADV), "Domestic Violence Facts," www.ncadv.org/files/DomesticViolenceFactSheet(National).pdf; and "Psychological Abuse," www.ncadv.org/files/PsychologicalAbuse.pdf.

[8]James J. Gross, "Emotion Regulation: Affective, Cognitive, and Social Consequences," *Psychophysiology* 39 (2002): 289.

[9]Jane M. Richards and James M. Gross, "Emotion Regulation and Memory: The Cognitive Costs of Keeping One's Cool," *Journal of Personality and Social Psychology* 79, no. 3 (2000): 410–424.

[10]James J. Gross, "Emotion Regulation: Affective, Cognitive, and Social Consequences," *Psychophysiology* 39 (2002): 281.

[11]Zindel Segal, interviewed by Miriam Boleyn-Fitzgerald, 6 August 2008, transcript of audio recording.

[12]John Teasdale, Zindel Segal, Mark Williams, Valerie Ridgeway, Judith Soulsby, and Mark Lau, "Prevention of Relapse/Recurrence in Major Depression by MBCT," *Journal of Consulting and Clinical Psychology* 68, no. 4 (2000): 615–623.

[13]Drs. Williams, Teasdale, Segal, and Kabat-Zinn outline the MBCT approach in their book, *The Mindful Way Through Depression: Freeing Yourself from Chronic Unhappiness* (New York: The Guilford Press, 2007).

[14]Zindel Segal, interviewed by Miriam Boleyn-Fitzgerald, 6 August 2008, transcript of audio recording.

[15]James J. Gross, "Emotion Regulation: Affective, Cognitive, and Social Consequences," *Psychophysiology* 39 (2002): 281–291.

[16]Zindel Segal, interviewed by Miriam Boleyn-Fitzgerald, 6 August 2008, transcript of audio recording.

[17]Naomi Law, "Scientists Probe Meditation Secrets," BBC News, 31 March 2008.

[18]National Institute of Mental Health, "Post-Traumatic Stress Disorder," 26 June 2008, www.nimh.nih.gov/health/publications/post-traumatic-stress-disorder-ptsd/index.shtml.

[19]Charles W. Hoge, Carl A. Castro, Stephen C. Messer, Dennis McGurk, Dave I. Cotting, and Robert L. Koffman, "Combat Duty in Iraq and Afghanistan, Mental Health Problems, and Barriers to Care," *New England Journal of Medicine* 351, no. 1 (1 July 2004): 13–22.

[20]Anthony P. King, interviewed by Miriam Boleyn-Fitzgerald, 29 January 2008, transcript of audio recording.

[21]Israel Liberzon, Anthony P. King, Jennifer C. Britton, K. Luan Phan, James L. Abelson, and Stephan F. Taylor, "Paralimbic and Medial Prefrontal Cortical Involvement in Neuroendocrine Responses to Traumatic Stimuli," *American Journal of Psychiatry* 164 (August 2007): 1250–1258.

[22]Richard Davidson, "Shaping Your Child's Brain," talk given at Appleton East High School (Appleton, Wisconsin), 13 May 2008.

[23]James A. Coan, Hillary S. Schaefer, and Richard J. Davidson, "Lending a Hand: Social Regulation of the Neural Response to Threat," *Psychological Science* 17, no. 12 (2006): 1037–8.

[24]Richard Davidson, interviewed by Miriam Boleyn-Fitzgerald, 26 June 2008, transcript of audio recording.

[25]Benedict Carey, "Lotus Therapy," *New York Times*, 27 May 2008.

[26]Aliza Weinrib, email communication, 20 October 2008.

[27]Richard Davidson, interviewed by Miriam Boleyn-Fitzgerald, 26 June 2008, transcript of audio recording.

Chapter 3

[1]Claudia Wallis, "The New Science of Happiness," *Time*, 17 January 2005.

[2]Ibid.

[3]Heather L. Urry, Jack B. Nitschke, Isa Dolski, Daren C. Jackson, Kim M. Dalton, Corrina J. Mueller, Melissa A. Rosenkranz, Carol D. Ryff, Burton H. Singer, and Richard J. Davidson, "Making a Life Worth Living: Neural Correlates of Well-Being," *Psychological Science* 15, no. 6: 367.

[4]Penelope Green, "This Is Your Brain on Happiness," *O Magazine*, March 2008.

[5]Richard Davidson, interviewed by Miriam Boleyn-Fitzgerald, 26 June 2008, transcript of audio recording.

[6]A. Quaranta, M. Siniscalchi, and G. Vallortigara, "Asymmetric Tail-Wagging Responses by Dogs to Different Emotive Stimuli," *Current Biology* 17, no. 6: R199–R201.

[7]Lauran Neergaard, "Brain's Reaction to Yummy Food May Predict Weight," Associated Press, 16 October 2008.

[8]Ibid.

[9]National Science Foundation (NSF) Press Release 08–199, "Bullies May Enjoy Seeing Others in Pain," 7 November 2008.

[10]J. Decety, K.J. Michalska, Y. Akitsuki, and B. Lahey, "Atypical Empathic Responses in Adolescents with Aggressive Conduct Disorder: A Functional MRI Investigation," *Biological Psychology* 80, no. 2 (February 2009): 210.

[11]Julie Steenhuysen, "Bullies May Get Kick Out of Seeing Others in Pain," Reuters, 7 November 2008.

[12]Jean Decety et al., "Atypical Empathic Responses in Adolescents with Aggressive Conduct Disorder: A Functional MRI Investigation," *Biological Psychology* 80, no. 2 (February 2009): 208.

[13]Radha Chitale, "Pain May Be Pleasurable for Some Bullies," ABC News, 7 November 2008.

[14]J. Decety, K.J. Michalska, Y. Akitsuki, "Who Caused the Pain? A Functional MRI Investigation of Empathy and Intentionality in Children, *Neuropsychologia* 46 (2008): 2607–2614.

[15]Claudia Wallis, "The New Science of Happiness," *Time*, 17 January 2005.

[16]Penelope Green, "This Is Your Brain on Happiness," *O Magazine*, March 2008.

[17]The Dalai Lama as quoted in Surya Das, *Buddha Is as Buddha Does: The Ten Original Practices for Enlightened Living* (New York: HarperOne, 2008), 37.

[18]Dian Land, "Study Shows Compassion Meditation Changes the Brain," University of Wisconsin-Madison News, 25 March 2008, www.news.wisc.edu/14944.

[19]Dan Rather, "Mind Science," *Dan Rather Reports*, HDNet, 8 April 2008.

[20]Ibid.

[21]A. Lutz, L.L. Greischar, N.B. Rawlings, M. Ricard, R.J. Davidson, "Long-term Meditators Self-induce High-amplitude Gamma Synchrony During Mental Practice," *Proceedings of the National Academy of Sciences* 101:16369–73.

[22]Dan Rather, "Mind Science," *Dan Rather Reports*, HDNet, 8 April 2008.

[23]Amanda Gardner, "Meditation Can Wish You Well, Study Says," *U.S. News and World Report*, 27 March 2008.

[24]Yongey Mingyur Rinpoche and Eric Swanson, *The Joy of Living: Unlocking the Secret and Science of Happiness* (New York: Harmony Books, 2007).

[25]Yongey Mingyur Rinpoche, "The Joy of Living: A Public Talk," given in Hartford, Connecticut, 9 August 2007, clip available at www.youtube.com/watch?v=m5bpe6fXuPk.

[26]Antoine Lutz, Julie Brefczynski-Lewis, Tom Johnstone, and Richard Davidson, "Regulation of the Neural Circuitry of Emotion by Compassion Meditation: Effects of Meditative Expertise," *PloS One* 3, no. 3 (March 2008).

[27]Clara Moskowitz, "Neuroscience May Explain the Dalai Lama," MSNBC's *LiveScience*, 27 March 2008.

[28]Ibid.

[29]Dan Rather, "Mind Science," *Dan Rather Reports*, HDNet, 8 April 2008.

[30]Richard Davidson, "Shaping Your Child's Brain," talk given at Appleton East High School (Appleton, Wisconsin), 13 May 2008, transcript of audio recording.

[31]Ibid.

[32]Ibid.

[33]Claudia Wallis, "The New Science of Happiness," *Time*, 17 January 2005.

[34]Richard Davidson, "Shaping Your Child's Brain," talk given at Appleton East High School (Appleton, Wisconsin), 13 May 2008, transcript of audio recording.

[35]Collaborative for Academic, Social, and Emotional Learning (CASEL), "The Benefits of School-Based Social and Emotional Learning Programs: Highlights from a Forthcoming CASEL Report," December 2007, www.casel.org/downloads/metaanalysissum.pdf.

[36]Daniel Goleman, "Some Big News About Learning," DanielGoleman.info, 15 February 2008, www.danielgoleman.info/blog/2008/02/15/some-big-news-about-learning/.

[37]Dan Rather, "Mind Science," *Dan Rather Reports*, HDNet, 8 April 2008.

Chapter 4

[1]National Institute on Drug Abuse, "Drugs, Brains, and Behavior: The Science of Addiction," April 2007, 18, www.drugabuse.gov/ScienceofAddiction/.

[2]National Institute on Drug Abuse, "Comorbidity: Addiction and Other Mental Illnesses," December 2008, 4, www.drugabuse.gov/researchreports/comorbidity/.

[3]National Institute on Drug Abuse, "Drugs, Brains, and Behavior: The Science of Addiction," April 2007, ii, www.drugabuse.gov/ScienceofAddiction/.

[4]Amanda Onion, "Chronic Pain Comes from the Brain," ABC News, 28 February 2005, www.abcnews.go.com/Health/PainManagement/story?id=531217&page=1.

[5]National Institute on Alcohol Abuse and Alcoholism, "Alcoholic Liver Disease," *Alcohol Alert* 64 (January 2005), www.pubs.niaaa.nih.gov/publications/aa64/aa64.htm

[6]A. Thomas McLellan, David C. Lewis, Charles P. O'Brien, Herbert D. Kleber, "Drug Dependence, a Chronic Medical Illness," *JAMA* 284 (2000): 1689–1695.

[7]National Institute on Drug Abuse, "Drugs, Brains, and Behavior: The Science of Addiction," 3, www.drugabuse.gov/ScienceofAddiction/. www.drugabuse.gov/ScienceofAddiction/.

[8]Jeneen Interlandi, "What Addicts Need," *Newsweek*, 3 March 2008.

[9]The American Pain Foundation, "Fast Facts about Pain," 2005, 1, www.painfoundation.org/page.asp?file=library/FastFacts.htm; and "Pain Facts: An Overview of American Pain Surveys," 2005, 9, www.painfoundation.org/page.asp?file=Newsroom/PainSurveys.htm.

[10]Amanda Onion, "Chronic Pain Comes from the Brain," ABC News, 28 February 2005, www.abcnews.go.com/Health/PainManagement/story?id=531217&page=1.

[11]Daniel Goleman, "Brain Images of Addiction in Action Show Its Neural Basis," *New York Times*, 13 August 1996.

[12]Substance Abuse and Mental Health Services Administration (SAMHSA), Office of Applied Studies, National Survey on Drug Use and Health, 2006 and 2007, www.oas.samhsa.gov/nsduh/reports.htm#Standard.

[13]Nora D. Volkow, Linda Chang, Gene-Jack Wang, Joanna S. Fowler, Yu-Sin Ding, Mark Sedler, Jean Logan, Dinko Franceschi, John Gatley, Robert Hitzemann, Andrew Gifford, Christopher Wong, and Naomi Pappas, "Low Level of Brain

Dopamine D$_2$ Receptors in Methamphetamine Abusers: Association With Metabolism in the Orbitofrontal Cortex," *American Journal of Psychiatry* 158 (2001): 2015.

[14]M.P. Paulus, N.E.Hozack, B.E. Zauscher, L. Frank, G.G. Brown, D.L. Braff, M.A. Schuckit, "Behavioral and Functional Neuroimaging Evidence for Prefrontal Dysfunction in Methamphetamine-Dependent Subjects," *Neuropsychopharmacology* 26, no. 1 (January 2002): 53.

[15]Nora D. Volkow, Linda Chang, Gene-Jack Wang, Joanna S. Fowler, Dinko Franceschi, Mark Sedler, Samuel J. Gatley, Eric Miller, Robert Hitzemann, Yu-Shin Ding, and Jean Logan, "Loss of Dopamine Transporters in Methamphetamine Abusers Recovers with Protracted Abstinence," *Journal of Neuroscience* 21, no. 23 (1 December 2001): 9414.

[16]WGBH/Frontline, *The Meth Epidemic*, 14 February 2006, www.pbs.org/wgbh/pages/frontline/meth/body/methbrainnoflash.html.

[17]Sandra Blakeslee, "This Is Your Brain on Meth," *New York Times*, 20 July 2004.

[18]American Cancer Society, "Cigarette Smoking," revised 14 November 2008, www.cancer.org/docroot/PED/content/PED_10_2X_Cigarette_Smoking.asp.

[19]Benedict Carey, "In a Clue to Addiction, Brain Injury Halts Smoking," *New York Times*, 26 January 2007.

[20]Ibid.

[21]Ibid.

[22]Jeneen Interlandi, "What Addicts Need," *Newsweek*, 3 March 2008.

[23]Ibid.

[24]Ibid.

[25]The American Pain Foundation, "Fast Facts about Pain," 2005, 1, www.painfoundation.org/page.asp?file=library/FastFacts.htm.

[26]A. Vania Apkarian, Yamaya Sosa, Sreepadma Sonty, Robert M. Levy, R. Norman Harden, Todd B. Parrish, and Darren R. Gitelman, "Chronic Back Pain Is Associated with Decreased Prefrontal and Thalamic Gray Matter Density," *Journal of Neuroscience* 24, no. 46 (17 November 2004): 10410.

[27]Ibid.

[28]Northwestern University, "Chronic Pain Harms the Brain," 5 February 2008, www.northwestern.edu/newscenter/stories/2008/02/chronicpain.html.

[29]Ibid.

[30]Melanie Thernstrom, "My Pain, My Brain," *New York Times*, 14 May 2006.

[31]R. Christopher deCharms, Fumiko Maeda, Gary H. Glover, David Ludlow, John M. Pauly, Deepak Soneji, John D. E. Gabrieli, and Sean C. Mackey, "Control Over - Brain Activation and Pain Learned by Using Real-time Functional MRI," *Proceedings of the National Academy of Sciences* 102, no. 51 (20 December 2005): 18629.

[32]Melanie Thernstrom, "My Pain, My Brain," *New York Times*, 14 May 2006.

[33]R. Christopher DeCharms et al., "Control Over Brain Activation and Pain Learned by Using Real-time Functional MRI," *Proceedings of the National Academy of Sciences* 102, no. 51 (20 December 2005): 18630.

[34]Ibid.

[35]Melanie Thernstrom, "My Pain, My Brain," *New York Times*, 14 May 2006.

[36]Emily Singer, "Looking at Your Brain on Drugs," *Technology Review*, 30 October 2006, www.technologyreview.com/Biotech/17674/.

[37]Ibid.

[38]Ibid.

Chapter 5

[1]Michael Koenigs, Liane Young, Ralph Adolphs, Daniel Tranel, Fiery Cushman, Marc Hauser, and Antonio Damasio, "Damage to the Prefrontal Cortex Increases Utilitarian Moral Judgments," *Nature* 446 (2007): 908-911.

[2]Jeffrey Rosen, "The Brain on the Stand," *New York Times*, 11 March 2007.

[3]Joshua Greene and Jonathan Cohen, "For the Law, Neuroscience Changes Nothing and Everything," *Philosophical Transactions of the Royal Society of London* 359 (2004): 1775.

[4]Peter Unger, *Living High and Letting Die: Our Illusion of Innocence* (New York: Oxford University Press, 1996).

[5]Joshua D. Greene, R. Brian Sommerville, Leigh E. Nystrom, John M. Darley, Jonathan D. Cohen, "An fMRI Investigation of Emotional Engagement in Moral Judgment," *Science* 293 (14 September 2001): 2106.

[6]Jeffrey Rosen, "The Brain on the Stand," *New York Times*, 11 March 2007.

[7]Joshua D. Greene, "Why Are VMPFC Patients More Utilitarian? A Dual-Process Theory of Moral Judgment," *Trends in Cognitive Sciences* 11, no. 8 (July 2007): 322.

[8]Michael Koenigs et al., "Damage to the Prefrontal Cortex Increases Utilitarian Moral Judgments," *Nature* 446 (2007): 908.

[9]Joshua Greene, "From Neural 'Is' to Moral 'Ought': What Are the Moral Implications of Neuroscientific Moral Psychology?" *Nature Reviews Neuroscience* 4 (October 2003): 850.

[10]*Roper v. Simmons* (03-633) 543 U.S. 551 (2005) 112 S. W. 3d 397, affirmed, www.law.cornell.edu/supct/html/03-633.ZS.html.

[11]Joseph T. McLaughlin, E. Joshua Rosenkranz, Timothy P. Wei, Stephane M. Clare, Aliya Haider, et al., "Brief of the American Medical Association, American Psychiatric Association, American Society for Adolescent Psychiatry, American Academy of Child and Adolescent Psychiatry, American Academy of Psychiatry and the Law, National Association of Social Workers, and National Mental Health Association as Amici Curiae in Support of the Respondent." *Roper v. Simmons*, U.S. Supreme Court, no. 03-633, www.abanet.org/crimjust/juvjus/simmons/ama.pdf.

[12]*Roper v. Simmons* (03-633) 543 U.S. 551 (2005) 112 S. W. 3d 397, affirmed, www.law.cornell.edu/supct/html/03-633.ZS.html.

[13]Joshua Greene and Jonathan Cohen, "For the Law, Neuroscience Changes Nothing and Everything," *Philosophical Transactions of the Royal Society of London* 359 (2004): 1775.

[14]Richard J. Davidson, Katherine M. Putnam, Christine L. Larson, "Dysfunction in the Neural Circuitry of Emotion Regulation—A Possible Prelude to Violence," *Science* 289 (28 July 2000): 594.

[15]Jeffrey Rosen, "The Brain on the Stand," *New York Times*, 11 March 2007.

[16]Henry T. Greely on the Law & Neuroscience Project, video recording for the MacArthur Foundation, 12 June 2008, www.youtube.com/watch?v=kveqkgYNIZs&feature=related.

[17]Stephen J. Morse on the Law & Neuroscience Project, video recording for the MacArthur Foundation, 12 June 2008, www.youtube.com/watch?v=L22lI4xTjXw&feature=channel.

[18]Michael S. Gazzaniga on the Law & Neuroscience Project, video recording for the MacArthur Foundation, 12 June 2008, www.youtube.com/watch?v=AOYws5Ok5Nk&feature=channel.

[19]National Science Foundation video, "Professor Michael Gazzaniga discusses the impact of neuroscience and the legal system," www.nsf.gov/discoveries/disc_videos.jsp?cntn_id=114979&media_id=65262&org=NSF.

[20]Michael S. Gazzaniga on the Law & Neuroscience Project, video recording for the MacArthur Foundation, 12 June 2008, www.youtube.com/watch?v=AOYws5Ok5Nk&feature=channel.

[21]Walter Sinnott-Armstrong on the Law & Neuroscience Project, video recording for the MacArthur Foundation, 12 June 2008, www.youtube.com/watch?v=av1EFK3QgsU.

[22]Owen D. Jones on the Law & Neuroscience Project, video recording for the MacArthur Foundation, 12 June 2008, www.youtube.com/watch?v=uKg-2fvJZKw&feature=channel.

[23]Michael Gazzaniga and Steve Mirsky, "Science Talk Podcast," *Scientific American*, 28 November 2007, www.scientificamerican.com/podcast/episode.cfm?id=82CE9C8B-E7F2-99DF-320EEC8640412E2D.

[24]Henry T. Greely on the Law & Neuroscience Project, video recording for the MacArthur Foundation, 12 June 2008, www.youtube.com/watch?v=kveqkgYNIZs&feature=related.

[25]Robert Lee Hotz, "The Brain, Your Honor, Will Take the Witness Stand," *Wall Street Journal*, 15 January 2009.

[26]Ibid.

[27]Walter Sinnott-Armstrong on the Law & Neuroscience Project, video recording for the MacArthur Foundation, 12 June 2008, www.youtube.com/watch?v=av1EFK3QgsU.

[28]Robert Lee Hotz, "The Brain, Your Honor, Will Take the Witness Stand," *Wall Street Journal*, 15 January 2009.

[29]Joshua W. Buckholtz, Christopher L. Asplund, Paul E. Dux, David H. Zald, John C. Gore, Owen D. Jones, and René Marois, "The Neural Correlates of Third-Party Punishment," *Neuron* 60, no. 5 (10 December 2008): 930-940.

[30]Robert Lee Hotz, "The Brain, Your Honor, Will Take the Witness Stand," *Wall Street Journal*, 15 January 2009.

[31]National Science Foundation video, "Professor Michael Gazzaniga discusses the impact of neuroscience and the legal system," www.nsf.gov/discoveries/disc_videos.jsp?cntn_id=114979&media_id=65262&org=NSF.

[32]Henry T. Greely on the Law & Neuroscience Project, video recording for the MacArthur Foundation, 12 June 2008, www.youtube.com/watch?v=kveqkgYNIZs&feature=related.

[33]John Tierney, "One Good Turn Deserves Another: Altruism Researchers Reply to Your Posts," *New York Times*, 26 June 2007.

[34]John Tierney, "Taxes a Pleasure? Check the Brain Scan," *New York Times*, 19 June 2007.

[35]William T. Harbaugh, Ulrich Mayr, and Daniel R. Burghart, "Neural Responses to Taxation and Voluntary Giving Reveal Motives for Charitable Donations," *Science* 316 (15 June 2007): 1623.

[36]Ibid, 1624.

[37]John Tierney, "One Good Turn Deserves Another: Altruism Researchers Reply to Your Posts," *New York Times*, 26 June 2007.

[38]Joshua Greene et al., "An fMRI Investigation of Emotional Engagement in Moral Judgment," *Science* 293 (14 September 2001): 2107.

[39]Peter Unger, *Living High and Letting Die: Our Illusion of Innocence* (New York: Oxford University Press, 1996).

[40]Joshua Greene, "From Neural 'Is' to Moral 'Ought': What Are the Moral Implications of Neuroscientific Moral Psychology?" *Nature Reviews Neuroscience* 4 (October 2003): 849.

[41]Ibid.

Chapter 6

[1]Lumosity website homepage, www.lumosity.com/.

[2]Nintendo's Brain Age website homepage, www.brainage.com/launch/index.jsp.

[3]R. Brookmeyer, E. Johnson, K. Ziegler-Graham, MH Arrighi, "Forecasting the Global Burden of Alzheimer's Disease," *Alzheimer's and Dementia* 3, no. 3 (July 2007): 186–91.

[4]Timothy Salthouse, "When Does Age-Related Cognitive Decline Begin?" *Neurobiology of Aging* 30, no. 4 (April 2009): 507–14.

[5]Howard Hughes Medical Institute, "Alzheimer's Disease Is Not Accelerated Aging," 30 September 2004, www.hhmi.org/news/buckner4.html.

[6]Randy L. Buckner, "Memory and Executive Function in Aging and AD: Multiple Factors that Cause Decline and Reserve Factors that Compensate," *Neuron* 44 (30 September 2004): 204.

[7]Howard Hughes Medical Institute, "Alzheimer's May Leave Some Forms of Memory Intact," 10 June 2004, www.hhmi.org/news/buckner3.html. See also Cindy Lustig and Randy L. Buckner, "Preserved Neural Correlates of Priming in Old Age and Dementia," *Neuron* 42 (10 June 2004): 865–75.

[8]Howard Hughes Medical Institute, "Training Improves Age-Related Memory Decline," 16 February 2002, www.hhmi.org/news/buckner2.html.

[9]Ibid.

[10]Howard Hughes Medical Institute, "Alzheimer's May Leave Some Forms of Memory Intact," 10 June 2004, www.hhmi.org/news/buckner3.html.

[11]Howard Hughes Medical Institute, "Brain Activity in Youth May Presage Alzheimer's Pathology," 24 August 2005, www.hhmi.org/news/buckner5.html.

[12]Ibid. See also Randy L. Buckner et al., "Molecular, Structural, and Functional Characterization of Alzheimer's Disease: Evidence for a Relationship Between Default Activity, Amyloid, and Memory," *Journal of Neuroscience* 25, no. 34 (August 24, 2005): 7709–17.

[13]Denise Grady, "Finding Alzheimer's Before a Mind Fails," *New York Times*, 26 December 2007.

[14]Roni Caryn Rabin, "Blood Sugar Linked to Memory Decline, Study Says," *New York Times*, 1 January 2009.

[15]Scott Small, interviewed by Miriam Boleyn-Fitzgerald, 15 July 2009, transcript of audio recording.

[16]Columbia University Medical Center, "Researchers at Columbia University Medical Center Link Blood Sugar to Normal Cognitive Aging," 30 December 2008, www.cumc.columbia.edu/news/press_releases/081230_Aging.html.

[17]Roni Caryn Rabin, "Blood Sugar Linked to Memory Decline, Study Says," *New York Times*, 1 January 2009.

[18]Scott Small, interviewed by Miriam Boleyn-Fitzgerald, 15 July 2009, transcript of audio recording.

[19]Roni Caryn Rabin, "Blood Sugar Linked to Memory Decline, Study Says," *New York Times*, 1 January 2009.

[20]Scott Small, interviewed by Miriam Boleyn-Fitzgerald, 15 July 2009, transcript of audio recording.

[21]"Staying Sharp: New Study Uncovers How People Maintain Cognitive Function In Old Age," *Science Daily*, 12 June 2009, www.sciencedaily.com/releases/2009/06/090608162424.htm.

[22]Ibid.

[23]M. Kindt, M. Soeter, B. Vervliet, "Beyond Extinction: Erasing Human Fear Responses and Preventing the Return of Fear," *Nature Neuroscience* (March 2009).

[24]Lesley Stahl, "The Memory Pill," CBS News *60 Minutes*, 17 June 2007. Video and transcript available at www.cbsnews.com/stories/2006/11/22/60minutes/main2205629_page2.shtml.

[25]Ibid.

[26]Robin Marantz Henig, "The Quest to Forget," *New York Times*, 4 April 2004.

[27]"Effect of Propranolol on Preventing Post-Traumatic Stress Disorder," NIH clinical trial no. NCT00158262, www.clinicaltrials.gov/ct2/show/NCT00158262.

[28]The President's Council on Bioethics, "Beyond Therapy: Biotechnology and the Pursuit of Happiness," Washington, D.C. (October 2003), 228, www.bioethics.gov/reports/beyondtherapy/.

[29]The President's Council on Bioethics, "Beyond Therapy: Biotechnology and the Pursuit of Happiness," Washington, D.C. (October 2003), 229, www.bioethics.gov/reports/beyondtherapy/.

[30]Robin Marantz Henig, "The Quest to Forget," *New York Times*, 4 April 2004.

[31]Lesley Stahl, "The Memory Pill," CBS News *60 Minutes*, 17 June 2007. Video and transcript available at www.cbsnews.com/stories/2006/11/22/60minutes/main2205629_page2.shtml.

[32]"Memories Selectively, Safely Erased in Mice," *Science Daily*, 23 October 2008, www.sciencedaily.com/releases/2008/10/081022135801.htm.

[33]Robin Marantz Henig, "The Quest to Forget," *New York Times*, 4 April 2004.

[34]Lesley Stahl, "The Memory Pill," CBS News *60 Minutes*, 17 June 2007. Video and transcript available at www.cbsnews.com/stories/2006/11/22/60minutes/main2205629_page2.shtml.

[35]Ibid.

[36]Joseph Z. Tsien et al., "Inducible and Selective Erasure of Memories in the Mouse Brain Via Chemical-Genetic Manipulation," *Neuron* 60, no. 2 (23 October 2008): 353–66.

[37]P. Serrano, E.L. Friedman, J. Kenney, S.M. Taubenfeld, J.M. Zimmerman, et al., "PKMzeta Maintains Spatial, Instrumental, and Classically Conditioned Long-Term Memories," *PLoS Biology* 6(12): e318.

[38]"Spotless Mind? Unwanted Memories Might Be Erasable Without Harming Other Brain Functions," *Science Daily*, 24 December 2008, www.sciencedaily.com/releases/2008/12/081223121137.htm.

[39]Benedict Carey, "Brain Researchers Open Door to Editing Memory," *New York Times*, 6 April 2009.

[40]Larry Rosenberg, *Living in the Light of Death: On the Art of Being Truly Alive* (Boston, MA: Shambhala Publications, 2000), 43–4.

Chapter 7

[1]"Cula-Malunkyovada Sutta: The Shorter Instructions to Malunkya" (MN 63), translated from the Pali by Thanissaro Bhikkhu, *Access to Insight*, 7 June 2009, www.accesstoinsight.org/tipitaka/mn/mn.063.than.html.

[2]Vilayanur S. Ramachandran, "A Journey to the Center of your Mind," *TED Talks*, March 2007, www.ted.com/index.php/talks/vilayanur_ramachandran_on_your_mind.html.

[3]Steven Pinker, "The Mystery of Human Consciousness," *Time*, 29 January 2007.

[4]Ibid.

[5]Cornelia Dean, "Science of the Soul? 'I Think, Therefore I Am' Is Losing Force," *New York Times*, 26 June 2007.

[6]B. Alan Wallace, interviewed at "Attention, Memory and the Mind: A Synergy of Psychological, Neuroscientific and Contemplative Perspectives," Mind and Life XVIII Conference, 9 April 2009, www.youtube.com/watch?v=8mKLN2bRtss.

[7]B. Alan Wallace, "The Conscious Universe," talk given at Unity Church in Santa Barbara, California, 16 January 2008, www.sbinstitute.com/LecturesMP3.html.

[8]Richard Davidson at "Attention, Memory and the Mind: A Synergy of Psychological, Neuroscientific and Contemplative Perspectives," Mind and Life XVIII Conference, 9 April 2009, as quoted on the Mind and Life Institute Blog, www.mindandlife.org/blog/2009/04/reflections-on-day-4-emotion-attention-memory/.

[9]William James, "The Consciousness of the Self," Chapter 10 in *Principles of Psychology*, Volume 1 (New York, Henry Holt and Co., 1890).

[10]Norman A. S. Farb, Zindel V. Segal, Helen Mayberg, Jim Bean, Deborah McKeon, Zainab Fatima, and Adam K. Anderson, "Attending to the Present: Mindfulness Meditation Reveals Distinct Neural Modes of Self-Reference," *Social Cognitive and Affective Neuroscience* 2, no. 4 (December 2007): 313–14.

[11]Zindel Segal, "Happiness and the Brain," *The Agenda with Steve Paikin*, TVO, 13 January 2009, www.tvo.org/TVO/WebObjects/TVO.woa?video?TAWSP_Dbt_20090113_779412_0.

[12]Norman A. S. Farb et al., "Attending to the Present: Mindfulness Meditation Reveals Distinct Neural Modes of Self-Reference," *Social Cognitive and Affective Neuroscience* 2, no. 4 (December 2007): 313.

[13]Ibid., 320

[14]Richard J. Davidson, "Well-Being and Affective Style: Neural Substrates and Biobehavioural Correlates," *Philosophical Transactions of the Royal Society* 359 (2004): 1395–1411.

[15]V.S. Ramachandran, talk given at *Beyond Belief: Science, Religion, Reason and Survival*, Salk Institute for Biological Studies, 5 November 2006, www.the-sciencenetwork.org/programs/beyond-belief-science-religion-reason-and-survival/session-4-1.

[16]"Split Brain Behavioral Experiments,"
www.youtube.com/watch?v=ZMLzP1VCANo&feature=player_embedded.

[17]Ibid.

[18]National Science Foundation video, "Professor Michael Gazzaniga discusses the impact of neuroscience and the legal system," www.nsf.gov/discoveries/disc_videos.jsp?cntn_id=114979&media_id=65262&org= NSF.

[19]V.S. Ramachandran, talk given at *Beyond Belief: Science, Religion, Reason and Survival*, Salk Institute for Biological Studies, 5 November 2006, www.the-sciencenetwork.org/programs/beyond-belief-science-religion-reason-and-survival/session-4-1.

[20]Ibid.

[21]Albert Einstein, Letter of 1950, as quoted in the *New York Times*, 29 March 1972.

[22]John T. Cacioppo and William Patrick, *Loneliness: Human Nature and the Need for Social Connection* (New York: W.W. Norton & Co., 2008).

[23]Debra A. Gusnard et al., "Medial Prefrontal Cortex and Self-Referential Mental Activity: Relation to a Default Mode of Brain Function," *PNAS* 98, no. 7 (27 March 2001): 4259–4264.

[24]Damien A. Fair, Alexander L. Cohen, Nico U. F. Dosenbach, Jessica A. Church, Francis M. Miezin, Deanna M. Barch, Marcus E. Raichle, Steven E. Petersen, and Bradley L. Schlaggar, "The Maturing Architecture of the Brain's Default Network," *PNAS* 105, no. 10 (11 March 2008): 4028–4032.

[25]Decety, J., Michalska, K.J., & Akitsuki, Y. "Who Caused the Pain? A Functional MRI Investigation of Empathy and Intentionality in Children," *Neuropsychologia* 46 (2008): 2607–2614.

[26]Sharon Begley, "Religion and the Brain," *Newsweek*, 7 May 2001.

[27]V.S. Ramachandran in "God and the Temporal Lobes," www.youtube.com/watch?v=qIiIsDIkDtg.

[28]V.S. Ramachandran, talk given at *Beyond Belief: Science, Religion, Reason and Survival*, Salk Institute for Biological Studies, 5 November 2006, www.the-sciencenetwork.org/programs/beyond-belief-science-religion-reason-and-survival/session-4-1.

[29]V.S. Ramachandran in "God and the Temporal Lobes," www.youtube.com/watch?v=qIiIsDIkDtg.

[30]V.S. Ramachandran, talk given at *Beyond Belief: Science, Religion, Reason and Survival*, Salk Institute for Biological Studies, 5 November 2006, www.the-sciencenetwork.org/programs/beyond-belief-science-religion-reason-and-survival/session-4-1.

[31]A. Chris Gajilan, "Are Humans Hardwired for Faith?" CNN.com, 5 April 2007, www.cnn.com/2007/HEALTH/04/04/neurotheology/.

[32]Sharon Begley, "Religion and the Brain," *Newsweek*, 7 May 2001.

[33]Richard Davidson, interviewed by Miriam Boleyn-Fitzgerald, 26 June 2008, transcript of audio recording.

[34]Bernard J. Baars, Thomas Z. Ramsoy, and Steven Laureys, "Brain, Conscious Experience, and the Observing Self," *Trends in Neurosciences* 26, no. 12 (December 2003): 671–675.

[35]Richard Davidson, interviewed by Miriam Boleyn-Fitzgerald, 26 June 2008, transcript of audio recording.

[36]Matthieu Ricard, "On the Habits of Happiness," *TED Talks*, February 2004, www.ted.com/index.php/talks/matthieu_ricard_on_the_habits_of_happiness.html.

[37]Jon Kabat-Zinn, "Reflections on the Origination, Development, and Scope of Mindfulness-Based Stress Reduction Programs in Mainstream Medicine," talk given at the Mind and Life XVI Conference, 16 April 2008, www.mindandlife.org/conf08. mayo.html.

[38]Yongey Mingyur Rinpoche, interviewed by Miriam Boleyn-Fitzgerald, 20 April 2009, transcript of audio recording.

[39]Penelope Green, "This Is Your Brain on Happiness," *O Magazine*, March 2008.

[40]Richard Davidson, interviewed by Miriam Boleyn-Fitzgerald, 26 June 2008, transcript of audio recording.

[41]Jon Kabat-Zinn, "Reflections on the Origination, Development, and Scope of Mindfulness-Based Stress Reduction Programs in Mainstream Medicine," talk given at the Mind and Life XVI Conference, 16 April 2008, www.mindandlife.org/conf08. mayo.html.

[42]Sakyong Mipham, *Turning the Mind into an Ally* (New York: Riverhead Books, 2003), 24.

[43]Ibid.

[44]Matthieu Ricard, "On the Habits of Happiness," *TED Talks*, February 2004, www.ted.com/index.php/talks/matthieu_ricard_on_the_habits_of_happiness.html.

[45]Sakyong Mipham, *Ruling Your World* (New York: Broadway Books, 2005), 30.

[46]Jack Kornfield, *A Path with Heart: A Guide Through the Perils and Promises of Spiritual Life* (New York: Bantam Books, 1993), 60.

[47]Larry Rosenberg, *Breath by Breath: The Liberating Practice of Insight Meditation* (Boston: Shambhala Publications, 1998), 20-1.

[48]Antoine Lutz, Heleen A. Slagter, John D. Dunne, and Richard J. Davidson, "Attention Regulation and Monitoring in Meditation," *Trends in Cognitive Sciences* 12, no. 4 (1 April 2008):164.

[49]Heleen A. Slagter, Antoine Lutz, Lawrence L. Greischar, Andrew D. Francis, Sander Nieuwenhuis, James M. Davis, Richard J. Davidson, "Mental Training Affects Distribution of Limited Brain Resources, *PLoS Biology* 5, no. 6 (June 2007): e138.

[50]Yongey Mingyur Rinpoche, *Joyful Wisdom: Embracing Change and Finding Freedom* (New York: Harmony Books, 2009), 146-7.

[51]Antoine Lutz et al., "Attention regulation and monitoring in meditation," *Trends in Cognitive Sciences* 12, no. 4 (1 April 2008):164.

[52]Mark Williams, John Teasdale, Zindel Segal, and Jon Kabat-Zinn, *The Mindful Way through Depression: Freeing Yourself from Chronic Unhappiness* (New York: The Guilford Press, 2007), 182.

[53]Ibid., 184.

[54]Ibid., 183.

[55]Pema Chodron, *The Wisdom of No Escape* (Boston, MA: Shambhala Publications, 1991), 4.

[56]Sharon Salzberg, "The Force of Kindness," public talk given at University of Wisconsin-Madison, 28 May 2009, transcript of audio recording.

[57]Sharon Salzberg, *Lovingkindness: The Revolutionary Art of Happiness* (Boston: Shambhala Publications, 2008), 19.

[58]Ibid., 21.

[59]Sharon Salzberg, "The Force of Kindness," public talk given at University of Wisconsin-Madison, 28 May 2009, transcript of audio recording.

[60]Jon Kabat-Zinn, "Reflections on the Origination, Development, and Scope of Mindfulness-Based Stress Reduction Programs in Mainstream Medicine," talk given at the Mind and Life XVI Conference, 16 April 2008, www.mindandlife.org/conf08.mayo.html.

[61]Matthieu Ricard, "On the Habits of Happiness," *TED Talks*, February 2004, www.ted.com/index.php/talks/matthieu_ricard_on_the_habits_of_happiness.html.

[62]Yongey Mingyur Rinpoche, interviewed by Miriam Boleyn-Fitzgerald, 20 April 2009.

[63]Vilayanur S. Ramachandran, "A Journey to the Center of your Mind," *TED Talks*, March 2007, www.ted.com/index.php/talks/vilayanur_ramachandran_on_your_mind.html.

Acknowledgments

My heartfelt gratitude to the book's first editor, Amanda Moran, for believing in the project and for nurturing it through its youngest stages; to my wonderful agent, Jodie Rhodes, for making the perfect match; to Tim Moore, Gina Kanouse, and Russ Hall for their steadfast support and thoughtful guidance from start to finish; to Jovana San Nicolas-Shirley and the production team at Pearson for handling the manuscript with care; and to the book's exceptional science editor, Kirk Jensen, for his insightful readings and spot-on advice.

I feel great appreciation for the teachers and researchers who took time out of their hectic lives to answer questions about their work, including Richard Davidson and his assistant Susan Jensen; Yongey Mingyur Rinpoche and his assistant Edwin Kelley; Zindel Segal, Anthony King, Scott Small, Adam Anderson, Norman Farb, and Aliza Weinrib; and for the researchers who so freely shared their remarkable images for reprint, including Steven Laureys (whose image is featured on the cover), Randy Buckner, Jean Decety, Christopher deCharms, Damien Fair, Joshua Greene, William Klunk, Terry Oakes, Adrian Owen, Nora Volkow, and Mark Wheeler.

I am indebted to everyone who taught me science and ethics and, most influentially, how to talk and write about them so that people might care. I thank Sarah Flynn, brilliant editor and true friend, who let me play sidekick and encouraged me to keep the pen moving. I am also deeply grateful to John Boccio, Diarmuid Maguire, Amy Bug, Barry Schwartz, Kevin Knobloch, Ruth Faden, Dan Guttman, Anna Mastroianni, Jeff Kahn, Gil Whittemore, Charlie Weiner, Tom Cochran, Stan Norris, Jacob Scherr, Harriet Ritvo, Hugh Gusterson, Leo Marx, Jill Conway, Deborah Fitzgerald, Evelyn Fox Keller, and especially to my parents—Alyce and Ken—who were the first people to teach me that science is for girls.

With love and admiration, I thank the family members who have sustained me from the get-go: my beautiful, strong, and creative contemporaries Elizabeth, April, Ken Jr., K.B., Steven, Chris, Peter, Heather, Greg, Pietro, Noah, and Martha; and our loving aunts and uncles—Tony, Polly, Lynn, and Frank—who continue to supply lifesaving quantities of unconditional acceptance and outstanding food. I also thank my second family, Barbara and Larry and their impressive clan—the Hacketts, Quinns, Fitzgeralds, and Dowds—who welcomed me from day one, in spite of my ill-mannered lack of Irish heritage.

My boundless thanks to the dear friends who've supported me on the way (carrying as necessary), in rough order of appearance: Connie Clark, Elizabeth DeFrancesco, Telory Davies, Matthew Fitzsimmons, Kathy Sturm-Ramirez, John Colaianni, Noel Theodosiou, Shamita Ray, Debby Holland, Diahanna Post, Victoria Reznik Krikorian, Kirsten Findell, Eric Hyett, Rami Ahmad, Tracy Iles-Leith, Charles Flanagan, Lydia and David Waite, Catherine Hollis, Joy Jordan, Mark Johnson, Karen Hoffmann, Kathy Privatt, Monica Rico, Peter Blitstein, Tanya Bunson, B. Alford, Meg and Dan Nickchen, Molly and J.P. Larson, Katie Tegge, Jen Kamm, Andrew Knudsen, Nancy Gates Madsen, Greg Madsen, Joshua Cobbs, Ben Rinehart, Michele Boge, Sarah Welch, and Sally Jaeger-Altekruse—with a special pride clap for Meg and Molly, who provided me with daily doses of encouragement and laughter, not to mention first-rate childcare on an as-needed basis.

It is with immense gratitude that I offer this book to my husband Patrick and my son Aidan, who fuel my writing process and my life with a potent blend of patience, unconditional love, and good humor. Aidan—who produces 100 books in the span of time it takes his mom to produce one (and, with age-inappropriate empathy, tries hard not to make me feel bad about it)—is my reason for everything. Patrick— who says "Yes" when I ask, "Can I borrow your eyes?" (and by "eyes," I mean his openhearted attention and clarity of mind, both of which he shares in unlimited quantities)—makes everything possible.

About the Author

For 15 years, **Miriam Boleyn-Fitzgerald** has published on a wide range of scientific topics geared to curious readers of all backgrounds. She believes in taking the boring out of "technical" so that we might all have ready access to knowledge that can help us lead happier, healthier, more fulfilling lives.

With a degree in physics from Swarthmore College, Ms. Boleyn-Fitzgerald is a former recipient of the Thomas J. Watson Fellowship and the Ida M. Green Award for graduate studies in Science, Technology, and Society at the Massachusetts Institute of Technology. She has worked as a staff writer for President Clinton's Advisory Committee on Human Radiation Experiments and as an analyst for the Natural Resources Defense Council and the Union of Concerned Scientists. Ms. Boleyn-Fitzgerald lives and writes in Appleton, Wisconsin, with her husband Patrick and their son Aidan.

Index

FT Press

FINANCIAL TIMES

In an increasingly competitive world, it is quality
of thinking that gives an edge—an idea that opens new
doors, a technique that solves a problem, or an insight
that simply helps make sense of it all.

We work with leading authors in the various arenas
of business and finance to bring cutting-edge thinking
and best-learning practices to a global market.

It is our goal to create world-class print publications
and electronic products that give readers
knowledge and understanding that can then be
applied, whether studying or at work.

To find out more about our business
products, you can visit us at www.ftpress.com.